Sea of Opportunity

Sea of Opportunity

The Japanese Pioneers of
the Fishing Industry in Hawai'i

Manako Ogawa

University of Hawai'i Press
HONOLULU

© 2015 University of Hawai‘i Press
All rights reserved
Printed in the United States of America

20 19 18 17 16 15   6 5 4 3 2 1

Library of Congress Cataloging-in-Publication Data
Ogawa, Manako, author.
  Sea of opportunity : the Japanese pioneers of the fishing industry in
 Hawai‘i / Manako Ogawa.
    pages   cm
  Includes bibliographical references and index.
  ISBN 978-0-8248-3961-1 (alk. paper)
  1. Japanese—Employment—Hawaii—History. 2. Fishers—Hawaii—
History. 3. Fishing—Hawaii—History. 4. Fisheries—Hawaii—History.
I. Title.
DU624.7.J3O373   2015
338.3'72709969—dc23   2014021615

University of Hawai‘i Press books are printed on
acid-free paper and meet the guidelines for permanence
and durability of the Council on Library Resources.

Designed by George Whipple

Printed by Sheridan Books, Inc.

Contents

List of Illustrations vii
Acknowledgments ix
Note on Names xiii

    Introduction *1*

1 Passage to Hawai'i: The Development of a Fishing Culture in Japan since Ancient Times *11*

2 Japanese Fisherman Enter Hawaiian Waters: The Formative Years of Commercial Fishing in Hawai'i and the Rise of the Japanese, from 1899 to the Early 1920s *33*

3 The Heyday of the Japanese Fishing Industry in Hawai'i *62*

4 Surviving the Dark Days *92*

5 The Reconstruction and Revitalization of Fisheries after World War II *116*

6 Okinawa and Hawai'i *129*

    Epilogue *153*

Notes 167
Bibliography 189
Index 199

Illustrations

Maps

Japan, circa 2010   *13*
Kishū (Wakayama Prefecture), circa 2012   *14*
Aki (Hiroshima Prefecture), circa 2012   *16*
Chōshū (Yamaguchi Prefecture), circa 2012   *17*
Hawaiian Islands   *35*
Honolulu, circa 1920   *46*
Oʻahu   *73*
Kauaʻi   *74*
Maui   *75*
Okinawa, circa 2012   *130*

Figures

Fishing boats on Waiakea River, Hilo, island of Hawaiʻi   *44*
Matsutarō Yamashiro   *47*
Japanese fishing boats at Pier 16, Honolulu Harbor, circa 1910–1920   *54*
Matsujirō Ōtani   *59*
Seiichi Funai   *67*
Skipjack tuna fishing before World War II   *69*
Skipjack tuna fishing before World War II   *70*
Kewalo Basin in 1935   *72*
Assembly lines at Hawaiian Tuna Packers in 1959   *76*
Hawaiian Tuna Packers   *77*
Japanese women on River Street in 1930   *79*

Japanese women peddling *81*
Ōtani at work *82*
Ebisu Shrine at Okikamuro *85*
Matsujirō and Kane Ōtani *103*
Inside Aʻala Marketplace *103*
Matsujirō Ōtani *118*
At the launching ceremony for a fishing boat *119*
Kenʼichi Nakabe *121*
Cadets at a sumo tournament *122*
At the launching ceremony for the *Kinan-maru* *125*
A female fish peddler purchases fish in Itoman *137*
Shintoku Miyagi relaxing on a bunk bed *139*
Hiroshi Nakashima and Tokusaburō Uehara *143*
Fishermen catching bait fish in shallow waters *145*
Skipjack tuna fishing during the early 1970s *146*
Lunchtime onboard *148*
Takeko Nakashima with her daughter, Lisa, and granddaughter, Laurally *159*
The Honolulu Fish Auction of the United Fishing Agency *161*
Ehime Maru memorial at Kakaʻako Waterfront Park *164*

Acknowledgments

I was a dedicated "landlubber" before I moved to Shimonoseki in 2006 to teach at National Fisheries University, the only institution of higher education in Japan specializing in fisheries. I believed that the sea was nothing but an unfathomable place that had little to do with my life. Although I always loved eating seafood, I barely thought of those who make their living on and from the sea. Daily interactions with my colleagues and students made me realize that the sea has always spawned unique cultures and societies. This book is the harvest of my wonderful days in Shimonoseki. In particular, I am very grateful to Natsuko Miki, Chitoshi Miwa, Kunio Nakashima, Nobuaki Itakura, Yūsuke Suda, and many other colleagues at National Fisheries University for directing my attention from the land to the sea and constantly offering valuable suggestions for improving my earlier drafts. I also express my special thanks to the students from fishing communities in northeastern Japan devastated by the 2011 Great East Japan Earthquake. Despite the tragic loss of their families and friends, their positive energy in rising from grief and rebuilding their hometowns taught me the resiliency and wisdom of sea people and inspired me to explore their history and culture.

I have also accumulated debts of gratitude to many other individuals and organizations. First of all, I would like to express my deep appreciation for Hiroshi Yoneyama for inviting me to his study group and securing a travel grant from Ritsumeikan University. Thanks to him, I was able to frequently visit Kyoto and debate with his feisty colleagues, Norifumi Kawahara, Mitsuhiro Sakaguchi, Masumi Izumi, Hiromi Monobe, Fuminori Minamikawa, and many other great scholars in the Kansai area. I also received academic support from outside Japan. I appreciate Sayuri Guthrie-Shimizu of Rice University for carefully reading my draft and giving me constructive suggestions. Masako Ikeda of the University of Hawai'i Press midwifed this book project by discovering the value of the fishing culture and urging me to add it to the literature on Japanese in Hawai'i. Tokiko Bazzell of the University of Hawai'i Hamilton Library generously shared her expertise on and experiences with the local Japanese and Okinawan communities. I also want to acknowledge other librarians and archivists at the sites I visited for research. The staff at the Museum of Japanese Emigration to Hawai'i in Suō-Ōshima, Okinawa Prefectural Archives, National Diet Library in Japan, and the Hawai'i State Archives were all knowledgeable and helpful.

From the birth of an idea for a book about Japanese fishing in Hawai'i to the completion of this manuscript, my fieldwork has been financially supported by National Fisheries University and Ritsumeikan University. It was also partially supported by the Japan Society for the Promotion of Science KAKENHI Grant Number 25243008. Portions of chapters 2 and 3 were published in the *Ritsumeikan Gengo Bunka Kenkyū* and *Hawaiian Journal of History;* part of chapter 4 appeared in *Chiiki Gyogyō Kenkyū;* and chapter 6 appeared in *Imin Kenkyū Nenpō.* I am grateful to the editors of these fine journals. I presented part of chapter 6 at the 2012 international symposium, "Remembering 40 Years since Reversion: Okinawan Studies until Now, Okinawan Studies from Now On," at Waseda University. I talked about part of chapter 4 at the 2014 annual conference of the Association for Asian American Studies in San Francisco. I appreciate the insightful comments from the audience that motivated me to improve my work. I also express my special thanks to the anonymous reviewers of my book drafts.

Many people in fishing communities gave generously of their time and knowledge. In Okikamuro, Shizuo Niiyama of Hakuseiji Temple willingly shared his precious collection of *Kamuro,* a monthly periodical published by the local youth of Okikamuro, and other valuable documents. Ryōko Ōtani searched for old pictures sent from her uncle, Matsujirō, and generously shared her old memories of the transpacific ties of her family. In Okinawa, Minami Kobashigawa and Hiromi Irei welcomed me like a family member by lodging me in their homes and serving me a feast of tasty Okinawan dishes. Yumiko Kashima of the Itoman city board of education provided me valuable local records of fisheries and introduced me to fishermen who had been to Hawai'i. Lucky Taxi Co. of Itoman allowed me to sit in its cozy waiting room and talk with taxi drivers, many of them having experiences of fishing in Hawai'i, while snacking on *sātāandāgī* (Okinawan doughnut balls). My interviews were pleasurable, thanks to Kaoru Shingaki, who accompanied me and not only interpreted strong Itoman accents into standard Japanese but also vividly told the stories of men and women of the Itoman district, her birthplace. Ken Uehara, a carpenter who makes traditional Okinawan *sabani* fishing boats, invited me to his studio and shared his knowledge and experiences of traditional Okinawan fishing. At Henza, Tamotsu Ashitomi and his wife, Masayo, searched their house for relics from Hawai'i at my request and found many interesting pictures. Moreover, local fish fresh from the sea served on the Ashitomis' dinner table pleased me greatly.

The fishing communities on the other side of the Pacific also shared great kindness and generosity. Akira Ōtani and his family made my research trip productive and pleasurable by showing me old records and pictures and treating me to lunch and dinner. Hiroshi Nakashima always surprised me by serving me huge Kona crabs he had caught. In addition, I would like to thank all of the people who willingly granted me interviews and shared their precious memories. Their names appear in the bibliography.

My largest debt belongs to my family. My father, Shin'ichirō Ogawa, and my mother, Kazuko Ogawa, have encouraged me to hold onto my dreams with all of my might. My husband, Yūichirō Taira, has always been an inspiration and my best colleague. I deeply appreciate him for tolerating my chronic absence and offering me crucial encouragement at various stages along the way. I dedicate this book to them.

Note on Names

In this book, the names of Japanese individuals are presented in the Western order, that is, with given name first and family name second. Exceptions to this order are reference citations of individuals who have published their writings in Japanese. In accord with contemporary standardization of the Hawaiian language, I use diacritical marks except where they did not appear in the original. I use the term "Native Hawaiian" to refer to the indigenous people of Hawai‘i, although "Hawaiian" is often used in the literature to refer to indigenous people in contemporary Hawai‘i.

Introduction

> If I stretched a line from the Gulf of Taiji, which I look at every day, to as far as the horizon, it would reach the California Coast. My father is there, and travels easily back and forth across the Pacific Ocean. As a child, I sometimes stumbled up to Tōmyō Point, following my grandfather. At Kandori Point, I often gathered shells and seaweed that were washed by the Black Current flowing from California. I never thought of America as a distant country.
>
> Ishigaki Ayako, *Ai to Wakare*

A common image of Japan is that the sea has isolated and separated it from the rest of the world. Those who have historically made a living by cultivating the land have often viewed the sea surrounding Japan as nothing more than an unfathomable mass with little direct relevance to their everyday lives. Indeed, many people believe that Japan has always been an agricultural country, characterized primarily by its rice production and consumption. However, the waters surrounding this island nation have also significantly contributed to shaping Japanese culture. Abundant seafood harvested from the ocean has always enriched the diet of the Japanese and has been deeply integrated into the various fabrics of Japanese society and culture. Even deep within mountain areas, farmers have customarily eaten dried sardines when they transplanted rice seedlings, and Shinto priests have never failed to use ocean fish, such as dried abalone and other sea creatures, in special rituals. Japanese people have treated sea salt, symbolizing seawater, as a special entity with purifying power, using it in a variety of ways at ceremonial times: when people come back from a funeral service, they sprinkle salt on their bodies to purify themselves; sumo wrestlers cast salt on the sumo ring before a match for the same reason. These widely practiced customs associated with the sea indicate that the Japanese have developed an extensive relationship with the surrounding waters for centuries, exploring the complex and even dangerous sea and shore, the length of which surpasses that of China. Japan is an archipelago of four major and about three thousand minor islands that extend approximately 1,875 miles from Okinawa to Hokkaido, "roughly the span of other major world fishing grounds, such as those found between the tip of Baja California and the mouth of the Columbia River, or between Miami and Nova Scotia."[1] The ocean currents sweeping along Japan's coasts—the warm northward-flowing Black Current (also known as Kuroshio) of the Pacific coast, the cold southward Okhotsk or Kurile Current (Oyashio) off northeastern Honshu, and the warm Tsushima Current, a branch of the Black Current flowing through the straits between Japan and Korea—create one of the most productive fishing regions in the world.

These geographical conditions have inspired Japan's people to nurture the characteristics of a maritime nation with a maritime destiny, actively exploring uncharted seas since ancient times.[2] During the third century, a Chinese scholar recorded the seemingly strange customs of the Japanese in *Gishi-wajin-den:* "all men, both old and young, had tattoos on their faces and bodies to scare off big fishes. . . . The fishermen of Wa (Japan) are good at diving and catch fishes and shells."[3] For a Chinese writer with a strong revulsion to swimming and diving, these activities were odd enough to be noteworthy. Regardless of the astonishment and ridicule expressed by their land-oriented neighbors, Japanese fishermen distinguished themselves by unhesitatingly exploring the sea in the pursuit of richly abundant aquatic resources. Some even dared to risk their lives to travel on the open seas in small boats when they were not satisfied to fish only along the shores. For numerous fishermen, the impulse to explore new fishing grounds was powerful enough to overcome their fears of violent seas and inspire them to range across the seas around and outside Japan, long before the advent of modern navigation and shipbuilding skills.

The appearance of Japanese fishermen exploring the different shores of the Pacific, including the Hawaiian archipelago, around the beginning of the twentieth century was, therefore, a natural consequence of the propensities that had developed over dozens of centuries. This book attempts to depict the Japanese as maritime people and to reveal their society and culture as central to the Japanese experience in the Hawaiian chain throughout the twentieth century. By moving Japanese fishermen from the periphery to the core of analysis, this work seeks to uncover their enterprise in Hawai'i, which the agricentric literature on the Japanese has often overlooked or obscured.

The history of Japanese Americans has captured the attention of historians in both the United States and Japan. Numerous scholars have studied Japanese immigrants by revealing their unilateral flow from Japan to Hawai'i and the continental United States, usually from "privation and disadvantage in the 'old country' to freedom and a future in a new land."[4] Through focusing on the process of assimilation and acculturation of Japanese into the mainstream of American society, they attempted to reveal a key metaphor enabling Japanese Americans to be integrated into the mainstream or the degree to which these people contributed to the diversification of American multiethnic society. In social and historical studies of Japanese Americans that depict ethnic contribution, or "success stories" as historian Franklin Odo calls them, the examination of ethnic identity has been the core.[5]

Unlike studies describing the transpacific movement of people as a one-way journey, some scholars have started looking more closely at the ongoing relationship between Japanese Americans, especially the issei (first generation), and their homeland, arguing the importance of their border-crossing flow not only from Japan to the United States, but also from the United States to Japan.[6] According to such studies, the Japanese American experience should be interpreted in the context of the multilateral travel of people, language, culture, and politics between the two coasts of the Pacific. These researchers have argued that the pattern of Japanese movement has been far more complicated and multilateral, and categorizing Japanese as issei as soon as they land on US territory and erasing their links to the homeland is hardly an appropriate conclusion.[7] Many Japanese in the United States retained important manifestations of their ethnic identities, and their strong attachment to Japan surfaced during specific moments in history, such as patriotic activities supporting Japan's continental expansion during the 1930s. It is, therefore, essential to locate the Japanese American experience in the transnational context of both west- and eastbound migratory flows.

This transnational approach is useful in arguing the diverse social, economic, and political aspects of Japanese communities in Hawai'i and North America. However, when most scholars of Japanese Americans use the term "transnational," the sea is disregarded in their investigations or is merely treated as the first and foremost obstacle to be transcended. This extremely agriculture-centered, land-bound assumption is appropriate for understanding Japanese workers in cane fields or on strawberry farms, but is misleading when applied to people who make their living on and from the sea. For fishermen who voyaged across the Pacific in search of new fishing grounds, the sea should never be interpreted as a hurdle to be cleared or a line dividing the lands. In their eyes, it was always a surface where various interests, skills, and powers would converge.

Seemingly because of the limited size of the fishing population compared to that of the farming population, people whose livelihoods depend on the sea have been marginalized in previous research regarding Japanese Americans in Hawaiʻi. But if we learn that their mobility and extent of travel overwhelmingly surpass those of the farmers, we may discover a new aspect of Japanese society that the agriculture-centered frameworks, the conventional modes of inquiry, can never reveal. Unlike farmers, who must focus exclusively on a particular section of land and devote themselves to it to improve its productivity, fishermen have to reduce their attachments to specific places and increase their geographical mobility to locate schools of fish, an elusive form of natural resource always moving about in the ocean. Thus, fishermen and farmers have completely different attitudes toward the land, even though fishing and farming tend to be thrown together as primary food production industries.

Over a period of centuries, Japanese fishing vessels operated in almost all parts of the Pacific Ocean with little awareness of any national borders. People of Yamaguchi and Hiroshima, for example, went through the Seto Inland Sea, a corridor connecting the Asian continent and the old capitals of Japan, Nara and Kyoto, and continued to sail the areas off the coasts of the Korean Peninsula and the East China Sea, as they had since ancient times. Even during the Tokugawa period (1603–1868), when the Tokugawa regime forbade Japanese to freely venture into foreign waters, people of the Seto Inland Sea developed the Kitamae ship lines, which covered areas around Japan, transgressing the boundaries set by several hundred feudal domains. Meanwhile, the people of Wakayama, who had been revered for their advanced fishing skills, penetrated the waters of Ezo (now Hokkaido), and even thrust into Sakhalin, whose jurisdiction, at that time, was disputed by Japan and Russia.

If we consider that the massive scale of these people's activity skillfully eroded this isolationism before the dawn of the Meiji era (1868–1912), it is natural that those with such nomadic propensities quickly fled the coast of Japan and spread throughout the Pacific Ocean as soon as the Meiji government opened the doors of Japan to the rest of the world. Wakayama started sending its people to Australia to catch pearl oysters and to Canada for salmon fishing even before the first government-contract immigrants, or *kanyaku imin,* reached Hawaiʻi in 1885. In the meantime, fishing boats from Yamaguchi, Hiroshima, and the prefectures of Kyushu immediately appeared off the shores of Korea, while fishermen from Okinawa gradually penetrated coastal areas of western Japan. Later, some of them headed for better fishing grounds far out in Southeast Asian waters. By the early decades of the twentieth century, fishermen from these areas had spread throughout various parts of the Pacific, including the coasts of Hawai'i, North, Central, and South America, Korea, China, Taiwan, Oceania, India, Southeast Asia, and the Pacific islands. They often moved around from one place to another for brief periods in search of better catches because they were not restricted to depleted fishing grounds or compelled to compete with rival boats. The Japanese from farming villages also spread throughout the Pacific regions and nations, but at a much slower pace than the fishermen.

Thus, people from fishing communities came to Hawai'i simply because the waters of the islands seemed rich enough to attract them. They developed *edamura* (branch villages) by inviting their family members, as they did in various places in and out of their homeland, to facilitate fishing. If Hawai'i ceased to be attractive, they would move without hesitation to another area. Following the migratory nature of the fish they sought, they constantly traveled around the ocean. The sojourner's mentality, a hotly debated topic in Japanese American society and history, was the norm among fishing societies. Indeed, many fishermen, especially those from Wakayama, left Hawai'i for the West Coast of the United States for better catches and more profitable business opportunities, while many of their counterparts from Yamaguchi chose to stay in Hawai'i and develop a modern conglomerate of fisheries. After World War II, hundreds of fishermen from Okinawa came to Hawai'i to supplement the dwindling population of local fishermen, but most eventually returned home. This pattern of life and movement is hard to find among farming communities.

Through decentering and recentering the colors and patterns of the land-based fabric, this study attempts to reconstruct Japanese lives from the sea while challenging the dominant analysis of Japanese in Hawai'i, which is confined primarily within the framework of the "cane culture."[8] Replacing the predominant image of Japanese as static farmers planting roots deep into the soil with the viewpoints of many Japanese fishermen, this work aims to reshape prior beliefs about the nature of Japanese society in Hawai'i and provide an alternative chronological history unique to fishing communities.

One of the primary purposes of this book is to reveal the leadership the Japanese brought to the fishing industries of Hawai'i beginning at the turn of the twentieth century. According to most accounts, the issei retained peasant status in Hawai'i's highly stratified society until the post–Pacific

War era, when the nisei (second-generation Japanese) became influential enough to assume leadership roles in the local community.⁹ When Japanese fishermen in the continental United States and Canada appear in historical studies, they are often portrayed as victims of white domination and targets of exclusion. In other words, they remained a minority group in the industry and occasionally fought against the limitations of fishing licenses and other regulations in order to protect their tiny shares of the white-dominated market.

In sharp contrast to these counterhegemonic efforts on the mainland, their counterparts in Hawai'i preserved their leading status, since they laid the foundation of modern commercial fisheries in the islands. It is true that most of the capital of the islands at the time was concentrated in the hands of a white oligarchy called the Big Five companies: Alexander and Baldwin, American Factors, Brewer and Co., Castle and Cooke, and Theo. H. Davies and Co. Their control began with the sugar industry and extended to dominance of the land, capital, the media, and government. The Big Five's power derived from their possession of abundant land and exploitation of cheap, unskilled labor from abroad. However, their domination did not infiltrate the fishing industry, on which the Japanese had a firm grip, and in which they never allowed the Big Five to meddle. Unlike the sugarcane business, where Japanese workers were in positions of "economic dependency,"¹⁰ Japanese on fishing vessels (*sampans*) established complete control of island waters and dominated the entire fishering process, from catching to processing and distribution of various marine products. The prerequisite of the fishing business was not the possession of vast amounts of land and the massive mobilization of unskilled workers, but rather advanced fishing skills and rich experience in handling highly perishable foods. The Japanese fishermen had these qualifications, and came to occupy an economic niche that the Big Five elites, the powers of the land, had failed to fill. By the early 1920s, the Japanese fishermen had developed the local fishing business into the third leading industry in the islands, behind sugar and pineapple production.

Gender analysis is also indispensable for a complete understanding of the industry. The fishing business in Japan could not function without the presence of women. Many women traveled on fishing boats with their husbands in waters around the nation, while others dove in search of shellfish, abalone, and agar weeds with the assistance of their husbands on boats. Statistical data show that only a handful of fisherwomen conducted business in Hawaiian waters, but many Japanese women played a crucial role in the industry, especially in the processing and distribution of their husbands' catch. These women's share of the work was derived from the customs and structures of fishing communities in Japan, where female members of fishing families often peddled their catch in farming villages or towns in exchange for rice, grain, and vegetables. Unlike people in agrarian communities, who could produce almost all of the necessities of life independently, people in fishing societies could not survive without trade. Fish is a commodity to be sold or bartered, and women played a pivotal role in connecting their society with the outside world. The sphere of women's peddling was, like that of their husbands' fishing, extensive.

With regard to economic activity, fundamental differences between farming and fishing villages produced different cultural values and beliefs in terms of gender relations and patterns of family life. For Japanese women who engaged in fishing and selling with great mobility, Hawai'i seems to have been a promising new market, and this assumption may have propelled them to come to Hawai'i along with their men, acting as their business partners. They joined their husbands in Hawai'i and North America for reasons other than being merely "picture brides," a symbol of patriarchal oppression.[11] Popular discourse of Japanese women depicts them as victims of intense patriarchy who sought refuge in the United States from the yoke of their male family members.

However, women of fishing villages encountered far fewer restrictions from men because the economic structure of their society valued individual skills rather than collective work, which undermined patriarchal authority. Even scholars who refute the "orientalist" description of Japanese culture tend to portray it as an exotic, timeless society bound by Confucianism and Bushido ethics, and adhere to the image of Japan's cultural orthodoxy within agricultural societies, where primogeniture was the norm and women's bodies and lives were under the complete control of their fathers and husbands. The land-based customs of inheritance and patriarchal practices, in which the perpetuation of the *ie* patriarchal household system was extremely important, were not applicable to the fishing communities. In addition, women's earnings through fishing and trading enabled them to achieve financial independence and autonomy. This is not to suggest that women in fishing villages enjoyed unlimited individual freedom from any form of family responsibility. It is simply to state that rendering any exclusive correlation between women's labor and patriarchal exploitation would be problematic. Researchers have asserted that Japanese women who migrated to the United States were freed from the restrictions placed upon them by the feudalistic *ie* system and gained a new opportunity to profit for themselves. According to their studies, these women's earnings in Japan were counted merely as a part of the family assets, whereas in the United States such women were productive units with access to social networks outside the *ie*.[12] This agriculture-centered, gross oversimplification of women's economic activities, exclusively in terms of a "collective" Japan and an "individualistic" America, overshadows the realities of the far more diverse and dynamic practices of Japanese women.

Focusing on the fishing culture, this book seeks to offer new insights into Japanese lives in Hawai'i. It interprets the arrival of Japanese fishermen and women in Hawaiian waters in the broader context of their extensive travel to various parts of the Pacific, and attempts to enrich the scholarship on the global expansion of fisheries in the Pacific Ocean, including whaling and fishing for salmon, tuna, cod, and many other sea creatures conducted by native Pacific islanders as well as Europeans and Americans. Recent historical studies on the Pacific have not only highlighted that the Pacific functioned as an indispensable space for commercial, cultural, and military exchanges of human beings, but also have demonstrated the more complicated historical relationship between humans and the ocean by including ocean animals as crucial constituents. Although the focus of this book remains on human activities and its geographical concern is restricted to Hawai'i, it can be integrated into the trend of rapidly expanding scholarship on the interactions among humans and between humans and sea creatures.[13]

Chapter 1 examines the history and culture of fishing in Japan, especially Wakayama, Yamaguchi, and Hiroshima prefectures, which sent many fishermen to Hawai'i who developed the modern commercial fishing industries there. Following the great national political upheavals that brought an end to the Tokugawa shogunate and began the Meiji Restoration in 1868, Japan changed from a feudal, isolated nation to the mightiest empire in Asia within a couple of decades. This rapid national transformation was accompanied by the impoverishment of ordinary people, which prompted them to emigrate to alleviate economic pressures. In addition to these factors, sea people were particularly motivated to go abroad, unlike farmers, and their willingness to venture to foreign waters was derived from a strong fishing tradition and culture that dated back to ancient times. The chapter distinguishes the historical, cultural, and social backgrounds of fishing communities from which a number of people traveled to Hawai'i and other areas of the Pacific Ocean.

Chapter 2 looks at the initial contact of Japanese fishermen with Native Hawaiians and the development of Japanese fishing industries in Hawai'i. Before the Japanese came to Hawai'i, Native Hawaiians had engaged in fishing to support their families and distributed their bounty among community members. At the beginning of the twentieth century, Japanese fishermen started operating in Hawaiian waters. Their efficient techniques instantly superseded those of their native rivals. Their success was not welcomed by the white-dominated local authorities and resulted in a series of anti-Japanese fishing regulations. The chapter begins by examining the early years of Japanese fishing in Hawaiian waters and subsequent interactions with other ethnic rivals. It then reveals how Japanese fishermen got along with the local hegemony and established themselves as a vital part of the Hawaiian fishing industry.

Chapter 3 treats the golden age of Japanese fishing, from the 1920s to the 1930s. During this period, almost all the fishing vessels in Hawaiian waters belonged to Japanese, and most of the workers in fish distribution and processing companies were Japanese, including many women. The growth of the industry was accompanied by the expansion of fishing communities, in which religious institutions were established to ensure the fishermen's safety at sea and well-developed social networks reduced the burdens of child care and household chores among the women. While the Japanese strengthened their position in the Hawaiian economy, the depletion of fishery resources off Hawaiian coasts due to overfishing and the nisei's unwillingness to continue in their parents' profession cast a shadow on the future of Japanese hegemony in the Hawaiian fishing industry. The chapter mainly discusses the social, cultural, and economic aspects of Japanese fishing communities and highlights how they managed to preserve their leadership throughout the 1920s and 1930s.

Chapter 4 deals with World War II. Japanese control of the fishing industry and their knowledge of the islands' beaches and coastlines triggered suspicions that Japanese fishermen were acting as agents of the Japanese empire. These suspicions became apparent among US military intelligence and high-level government officials, including the president of the United States. Alleged connections between Japanese fishermen and the imperial expansion of their homeland, together with the worsening relationship between Washington and Tokyo over Japan's continental aggression, resulted in new fishing regulations targeting the Japanese and the impounding of fishing boats for alleged violations of the rules of registering alien ships. Immediately after Japan's attack on Pearl Harbor,

Japanese fishermen at sea were strafed by American airplanes. Japanese leaders in Hawai'i's fishing industry were interned, and fishermen of Japanese ancestry were forbidden to go to sea. The scale of business in the Japanese fishing industry was drastically reduced under martial law. Nevertheless, anti-Japanese measures by military authorities did not completely eliminate the Japanese from island waters. The chapter examines the impact of the war on the industry as a whole and reveals how the Japanese managed to survive those harsh years and lay the groundwork for the postwar resurgence of the fishing industry in Hawai'i.

Chapters 5 and 6 explore the postwar reconstruction and restructuring of the Hawaiian fishing industry. Issei leaders returned from internment and started working to regenerate the business. However, they soon confronted formidable problems: the advanced age of former issei fishing boat crewmembers hindered them from going back to sea, while many of their nisei children were reluctant to replace their aging fathers in the industry. On the other hand, the demand for fresh fish, the availability of which was severely limited during the war, was strong after the war. Thanks to the wartime suspension of fishing in Hawai'i's waters, fish were even more abundant than before the war, providing a lucrative supply for the comeback of the fishing industry in postwar Hawai'i. This situation encouraged the Japanese leaders of the industry to approach the Japanese and Ryukyu governments, convincing the latter to send experienced Okinawan fishermen to Hawai'i. In addition, the shortage of Japanese workers resulted in the aggressive recruitment of people of other ethnicities to work on fishing boats. The fishing industry in Hawai'i revitalized itself by fundamentally transforming the previous Japanese domination into a multiethnic enterprise.

The epilogue highlights the ethnic diversification of the islands' fishing industry, which catered to the drastically changing economic, ethnic, social, and cultural structures of twenty-first-century Hawai'i. Although the Japanese fishing population has shrunk dramatically, there undoubtedly is some continuity in fishing methods, fish peddling, fish consumption, and many other aspects of maritime culture that the Japanese introduced more than a century ago. The chapter indicates that Japanese fishing traditions continue to evolve in Hawai'i amid a surge of transitions and transformations.

1 Passage to Hawai'i

The Development of a Fishing Culture in Japan since Ancient Times

Weaving a network of commercial fishing enterprises in Wakayama, Hiroshima, and Yamaguchi before the modern era

Kishū, what is now Wakayama and the southernmost part of Mie prefectures, was a bow-shaped region on the southern end of Honshu Island. Its long coastline directly facing the Pacific Ocean has long had abundant fish brought by the Kuroshio Black Current, coming from the Philippines and Indonesia in the summer. The abundance of maritime resources has attracted fishermen, called *kaifu,* since ancient times. *Kaifu,* under the leadership of the professional fishermen's clan, called Azumi-no-muraji, had a special duty to provide marine products to the imperial court. Their outstanding fishing activities earned widespread fame for Kishū as a region possessing the most advanced fishing skills in Japan. Kishū was, however, hostile to agriculture, with distinct geographical features such as lush mountains dropping sharply into the sea, discouraging people from cultivating plants or crops. In contrast to the scarcity of flat, arable land, the Black Current washing along the coast brought rich aquatic resources, in particular, skipjack tuna, whales, sardines, turban snails, and agar seaweed. These topographical features inspired the people of Kishū to seek their fortunes in the sea rather than in agriculture.

The abundant fruits of the sea constituted only half of the essential prerequisites for prosperity in commercial fishing; proximity to a consumer community was also indispensable for the development of fishing as an industry. Due to an absence of consumers, many solitary islands in the middle of the ocean lost their commercial fishing industries. In contrast, Kishū satisfied both conditions, because it was near the old capital cities of Nara, Kyoto, and Osaka and their great consumer populations. Moreover, the old capitals and their outskirts, known as the Kinai Plain, where commercial agriculture developed early, needed large amounts of fertilizer, which could be made from dried sardines. In order to feed the large population, the development of intensive farming and remarkable advances in agricultural skills in this region, in turn, produced a growing demand for effective fish fertilizers. Constant pressure from an expanding market encouraged the fishermen of Kishū to move beyond the rudimentary stages of beach seine fishing into far more sophisticated and efficient styles of boat seining. They pioneered the technique of casting fishing nets from boats earlier than any other place in Japan.[1]

Together with the innovation of new fishing nets and techniques, the improvement of navigational expertise encouraged the people of Kishū to venture into distant waters while spreading their advanced fishing methods throughout the nation. Many fishermen from the northwest coast of Kishū, for instance, had gone to the Kanto area and contributed to the development of commercial fishing since the dawn of the Tokugawa period (1603–1868). The rapidly expanding population of Kanto as a new capital under shogunate rule sparked the cultivation of grains, cotton, indigo plants, vegetables, and fruits, including burdocks, eggplants, radishes, and tangerine oranges. As a result, farmers persistently sought the fish fertilizer that Kishū fishermen supplied through their constant exploration of new fishing spots and improvements in equipment. Their active operations off the coast of the Bōsō Peninsula soon transformed it into one of the major fishing regions in Japan.

While some fishermen from Kishū settled in Kanto and concentrated on developing the local fisheries, others kept moving northward, searching for more productive waters. Fishing is akin to hunting, and these hunters of the sea tenaciously pursued better hauls by venturing into incredibly rough, but fertile, seas off the Oshika Peninsula, Kamaishi, and Ezo (Hokkaido). By the late Tokugawa period, Kishū fishermen had reached the sea surrounding southern Sakhalin and the Chishima (Kuril) Islands to fish for salmon and herring. They also began selling herring, salmon, and sea tangles to the aboriginal Ainu people of Ezo.[2] Later, they extended their business to include lumber, mining, and salmon hatching.

The intrusion of the Kishū fishermen into Ezo and Sakhalin, both of which were sparsely inhabited by the Japanese, assumed significance beyond private business enterprise, because, at that time, the territorial rights

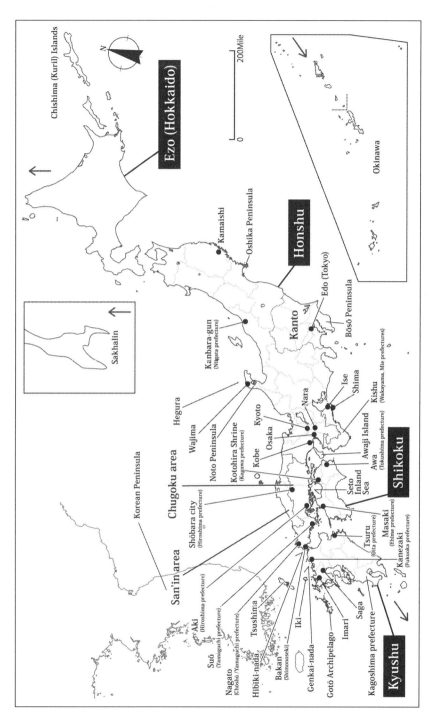

Japan, circa 2010.

Kishū, showing the contemporary municipalities of Wakayama Prefecture, circa 2012. Kushimoto was a part of Nishi-muro County until 2005.

to these lands were unsettled between Japan and Russia. In 1799, the Tokugawa shogunate placed Sakhalin and the eastern half of Ezo under its direct control, although the geographical contours of the former remained uncharted until 1809, when Rinzō Mamiya did a detailed survey and revealed that it was an island. The uncertain sovereignty of the northern land made it vulnerable to Russia's southward drive. This was manifested by the occasional appearance of Russian fleets at Ezo, their frequent contact and trade with the Ainu, and repeated requests to the Tokugawa regime to open up trade. At this critical juncture, Japanese authorities expected the Kishū entrepreneurs operating extensively in Ezo to arm themselves and play a vital part in border defenses by according them samurai warrior status. This transfer of Japan's national defense to private hands revealed the enfeebled powers and authority of the Tokugawa shogunate as well as the prosperity and experience accumulated by the pioneering enterprises of the Kishū fishermen's descendants.³

Besides the eastern- and-northern-bound ventures to Kanto and beyond, Kishū also sent fishing fleets to the Seto Inland Sea, the main artery of marine transportation from the old capital cities to Kyushu. The position of the numerous islands relative to one another produced complex currents with varying speed and strength in the Seto Inland Sea. These natural conditions sheltered various kinds of fish, and the coasts were dotted with prosperous castle towns with active commerce that hosted the Kitamae sailing ships, which were active in the shipping business and connected cities and towns from Ezo to Osaka. The abundance of marine resources and the large consumer population of Seto stimulated fishing enterprises among coastal communities.

The westward advance of Kishū fishermen was triggered by the transfer of domain lord Nagaakira Asano from Kishū to Aki (now the western part of Hiroshima Prefecture) in 1619. This historical event had a significant impact on the fisheries of the Seto Inland Sea, because Kishū sent many fishermen to Aki with greatly improved fishing nets. The appearance of the Kishū fishermen with their highly effective fishing skills and paraphernalia instantly overwhelmed their local rivals, forcing the latter to petition Lord Asano to send the newcomers back to Kishū. However, after the initial friction and confusion were overcome, the direct contact between the people of Kishū and Aki resulted in the notable expansion of fishing industries in Hiroshima and neighboring regions. The influence of Kishū was especially noticeable in areas without the long beaches necessary for drawing a beach seine because the new fishnet was thrown out from boats. Moreover, the merging of instruments from both places sparked the creation of a new fishing net, called *utase-ami*, suitable for Hiroshima Bay, the floor of which was predominantly sand. Fishing boats pulling *utase-ami* soon became well integrated with the seascape of Hiroshima Bay.

The introduction of the latest fishing net was not, however, a suitable solution for those who fished in straits or channels characterized by swift ocean currents on the surface and sunken rocks of various sizes on the bottom. Such places were ideal habitats for many kinds of fish, including sea bream, but highly unfit for the use of fishnets and long lines because both could be easily snagged or washed away. The application of a new material called *tegusu* to pole-and-line fishing provided a chance to venture to previously unexplored straits and channels. This revolutionary material was highly transparent and remarkably strong. Originally, it was

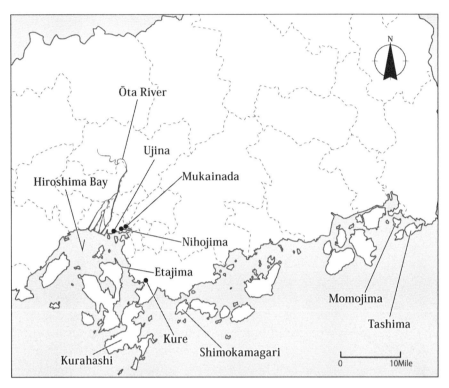

Aki, showing the contemporary coastline of Hiroshima Prefecture, circa 2012. Land reclamation, which has continued for centuries, has narrowed Hiroshima Bay. Nihojima and Ujina, both connected to the mainland, were islands during the Tokugawa period (1603–1868).

used as a wrapping string for medicine imported from China. Fishing communities in Kishū, Awaji Island, and eastern Awa (now Tokushima Prefecture), all of which faced straits with strong currents dividing Honshu, Awaji, and Shikoku islands, recognized its possible applications to fishing. Thanks to *tegusu*, they succeeded in dramatically enlarging their catches by pole-and-line fishing. Some of them were not satisfied merely to improve their haul, and thought about a new business; they loaded fishing boats with *tegusu* lines and poles and started selling their products at fishing villages on the coast of the Seto Inland Sea.

The appearance of the *tegusu* triggered the rapid development of *ipponzuri* (pole-and-line fishing) in Okikamuro, a small, mountainous island measuring only about 232 acres, lying off the coast of Suō-Ōshima Island in Suō (now the eastern part of Yamaguchi Prefecture). Despite its location in the middle of the Chugoku area and the island of Shikoku, it had functioned as a transit port where sailboats waited for better winds and currents. It is, therefore, wrong to characterize Okikamuro as a lonely, isolated island. The Hakuseiji temple of Okikamuro was used for lodging by feudal loads who sailed to and from Osaka during the Tokugawa period. Thus, Okikamuro played a crucial role in maritime trade and transportation in the Seto Inland Sea. In contrast, it did not fully develop commercial fishing, because it did not have beaches necessary for drawing a beach seine. What made matters worse, the discriminatory practices in Chōshū (now Yamaguchi Prefecture) required Okikamuro fishermen to use their nets elsewhere. Chōshū classified five villages of its domain as *otate-ura* (literally, a special standing beach) and granted them fishing privileges in exchange for imposing marine transportation duties for both the domain and shogunate officials and criminals. The fishermen of *otate-ura* also had to serve as a

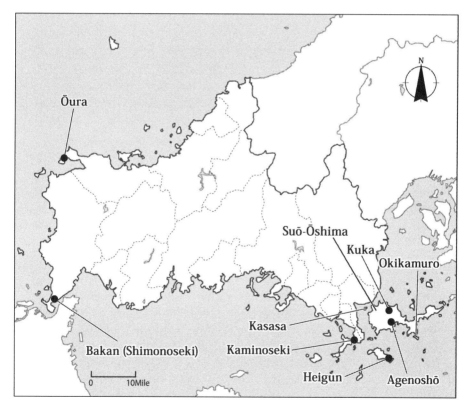

Chōshū (Yamaguchi Prefecture), circa 2012.

navy in case of war. Because Okikamuro was not one of the five *otate-ura* villages, its people could not freely use fishing nets.

However, the introduction of *tegusu* liberated them from feudal limitations, because pole-and-line fishing was beyond the control of the *otate-ura* system and they could engage in it freely. Moreover, high-priced sea bream, yellowtail, flatfish, and sea bass were available in abundance off the coast of Okikamuro. Unlike net fishing, which required great capital and the mobilization of a large workforce, a fisherman could start pole-and-line fishing with a small boat and a fishing rod. The fisherman could catch valuable fish alive and immediately sell them to merchants waiting on nearby trading ships. These ships were equipped with wells with flowing seawater, enabling the merchants to carry live fish to markets in Osaka and other cities of the Seto Inland Sea. Live sea bream and other fish from Okikamuro sold at higher prices among affluent customers. Not long after the introduction of *tegusu,* Okikamuro established itself as a prominent location for pole-and-line fishing.[4] Such stories of fisheries in Wakayama, Hiroshima, and Okikamuro reveal that people from Kishū and the Seto Inland Sea had already met and woven a network of commercial fishing skills and enterprises long before they encountered each other in Hawaiian waters at the turn of the twentieth century.

The Japanese from various parts of Kyushu, as well as the Seto Inland Sea, had advanced to the coast of the Korean Peninsula in the medieval period. In the fifteenth century, the king of Korea officially opened three ports to the Japanese and allowed them to build towns and conduct commercial fishing. The old records show that the Japanese population included both men and women; while men caught fish, women traded their catch for rice to eat and cotton to export to Japan.[5] Some Japanese in Korea embarked on export and import businesses. One of their sale items was salt gathered in the Seto Inland Sea.[6]

The presence of Japanese fishermen in Korea peaked during the early Tokugawa period, and Tsushima, an island situated between Kyushu and the Korean Peninsula, functioned as a transit point for those crossing the Tsushima and Korean straits. Because Tsushima itself had an extremely small population of full-time fishermen and its fishing industry was half developed, people from other places came to Tsushima and advanced into Korean waters.[7] An end was officially put to these enterprises, however, by a seclusion policy enforced by the Tokugawa shogunate in the 1630s. Nevertheless, the rich waters of Korea never ceased to attract Japanese, who sneaked in and poached fish. Among them were the fishermen of Aki. The marriage between the daughter of Lord Asano, of Aki, and Lord Sō, of

Tsushima, during the early days of the Bunka era (1804–1817) provided them with a permit to go to Tsushima.

This new privilege was a blessing for Aki, and especially for the Nihojima and Mukainada areas, both of which were located at the narrow end of the Ōta River delta facing Hiroshima Bay. The residents of these areas made their living through fishing and farming, but ongoing land-reclamation work since the Tokugawa period had made their narrow fishing grounds even narrower. These geographical changes motivated them to disperse to various parts of the Seto Inland Sea, and some of them even plowed through the rough waves of Genkai-nada to Tsushima. When they appeared in Tsushima, they immediately started pole-and-line fishing and long-lining to catch squid, sea bream, and yellowtail. Moreover, some fishermen from Aki joined those from Nagato, Saga, and the island of Iki, and advanced into Korean waters chasing sardines and mackerel, tacitly violating the isolation policy of the Tokugawa shogunate.[8] In the meantime, some Aki fishermen went to the Gotō Archipelago to join whaling fleets. This archipelago and surrounding waters, called Saikai (Western Sea), were on a whale migration route, which originally drew fishermen from Kishū with advanced whaling skills and later attracted those from the Gulf of Hiroshima and its vicinities.

The expedition of Aki fishermen to Tsushima and beyond encouraged their neighbors from Suō-Ōshima and Okikamuro to follow in their wake. Fishermen from Kuka, of Suō-Ōshima, one of the *otate-ura* villages, and Okikamuro had already forged into the waters of Hibiki-nada, Genkai-nada, the Gotō Archipelago, and the island of Iki by the mid-Tokugawa period, but it was not until the last years of the Tokugawa regime that they sailed into Tsushima and started pole-and-line fishing to catch yellowtail, mackerel, and sea bream.[9] The advancement of Japanese to Tsushima and Korea via Genkai-nada indicates they had established a wide network of fishing culture from the Seto Inland Sea, north and west Kyushu, the San'in area, and Tsushima, to the southeastern coast of the Korean Peninsula by the end of the Tokugawa era. The official opening of Japan in 1858 through the port treaty system sparked their full-scale fishing expeditions to the foreign waters.

The opening of Japan and the expansion of long-distance voyaging

The official discontinuation of the seclusion policy at the end of the Tokugawa period reintegrated Japan into the world community, after which it was buffeted by a rough wave of Western imperialism. The new

Meiji government, which had toppled the Tokugawa shogunate in 1868, promoted the modernization and militarization of the state at a rapid pace in order to catch up to the Western powers and fend off the imperialist intrusion. The great upheavals of the nation caused by the new national policy, *fukoku-kyōhei* (enrich the nation and strengthen the military), had a significant impact on the lives of coastal communities.[10] During the Tokugawa period, fishing enterprises were conducted according to the rule that anyone could freely fish offshore, while only tenants of the land could catch fish onshore. This fundamental principle was, however, abolished by the Meiji government, which declared state ownership of the sea in 1875. The government divided the sea up into squares and leased each one to those who wished to fish.

This new regulation immediately caused considerable confusion among fishing villages and pitted them against each other over fishery rights, because it was impossible to strictly demarcate the sea surface. Moreover, the dividing lines did not necessarily correspond to the habitats and natural movement of fish. Surprised at the confusion and inefficiency the regulation caused, the government rescinded it one year later. However, it was too late to restore the previous order. Besides, the serious deflation caused by a belt-tightening policy under the leadership of Finance Minister Masayoshi Matsukata reduced the market for marine products and drove fishermen into a tighter corner.

The rapid military buildup also had adverse effects on fishing societies. The massive-scale land reclamation in Hiroshima Bay and the subsequent construction of Ujina military port diminished the fishing grounds of people in Nihojima and other fishing villages; the construction caused ecological disaster, devastating the oyster and seaweed farms floating on the shallow waters. The transformation of Kure into a naval base and the transfer of the naval academy from Tsukiji in Tokyo to the island of Etajima at the entrance of Hiroshima Bay set further limits on civilian use of these vicinities. In addition to the loss of fishing grounds in the bay, the drastic social, cultural, and economic changes following the destruction of the feudal system distressed fishing societies, especially those of the *buraku*. During the Tokugawa period, *buraku* residents were considered at the bottom of the social hierarchy and faced serious discrimination. However, this social ostracism had the positive side effect of protecting their exclusive engagement in so-called taboo work, according to Buddhism, such as the disposal of dead cows and pigs and making leather. Yet the disintegration of the old status system and the liberalization of occupation deprived the *buraku* of their monopoly in the leather industry and related jobs. In the meantime, *buraku* fishing communities were mercilessly deprived of their fishing spots by military institutions, and persistent discrimination made it impossible for them to obtain alternate fishing sites. They were the first to lose their rights and the last to recover them.

The diminishing means for the *buraku* to make a living pressed the sea and shore *buraku* fishermen to venture into distant waters. Around 1884, a *buraku* district in Hiroshima Prefecture began sending out fishing convoys to Korea. In the following year, the fishermen succeeded in processing and dispatching dried sardines to Japan. The people of Nihojima, Mukainada, Kure, Shimokamagari Island, and Kurahashi Island, all of which were located within Hiroshima Bay or its vicinity, followed them and started fish netting sardines.[11] The Korean waters were well known among the people of Aki, or Hiroshima Prefecture, after the abolition of feudal domains, because their ancestors had appeared there long before the dawn of the Meiji era (1868–1912).

The waves of modernization reached the coast of Okikamuro, which also suffered from intensifying disputes over fishing rights with its neighbors when new regulations limited access to previously open waters. The loss of fishing sites urged the people of Okikamuro to travel to distant waters. By 1878, more than one thousand people, or one-third of the entire population of Okikamuro, were constantly out in the seas of other prefectures. Less than a hundred people, most of whom were young novice fishermen or the half-retired elderly, remained on the island, but never hesitated to sail out to fish in the sea off Takamatsu in Shikoku. This amazing propensity for deep-water fishing soon evolved into the formation of the long-distance voyaging fleets Bakan-gumi ("gumi" means group) and Imari-gumi, the former headed for Bakan (Shimonoseki) to fish primarily at Hibiki-nada and Genkai-nada, whereas the latter went to the islands of Iki and Tsushima, and the Gotō Archipelago.

The destinations of fishing expeditions from Okikamuro did not stop at the Tsushima Strait. Two fishermen of Imari-gumi, Kanjirō Hara and Tatsunosuke Nakayama, went to Korea on the recommendation of Nagasaki fishermen living in Tsushima. The sea of Korea was shallow and rich in maritime resources, and one haul paid 2,800 yen, an extraordinary amount in 1879. Surprised at this achievement, their colleagues immediately formed a new fleet named Chōsen (Korea)-gumi and rushed to Korea. Thereafter, the number of Chōsen-gumi rapidly increased.[12] In 1902, the number of its fishing boats reached thirty-five, with 210 crews. They primarily engaged in pole-and-line fishing to catch sea bream, shark, and mackerel, but sometimes used longlines as well.[13]

In Korean waters, the Okikamuro fishermen met many Japanese rivals from Hiroshima Prefecture, Kyushu, and other places. Since the early Meiji period, going to Korea had become popular throughout the fishing villages of western Japan, despite the absence of official agreement on fisheries between the Japanese and Korean governments until 1883. Unlike the waters of Japan, which were depleted due to overfishing and overcrowding, Korea was full of many unexplored fishing spots blessed with sardines, sharks, sea bream, anglerfish, squid, Spanish mackerel, and other aquatic resources. In 1892, 683 Japanese fishing boats went to Korea and operated there. Of them, Hiroshima Prefecture dispatched 270 ships, while Yamaguchi Prefecture, including Okikamuro, followed with 155. Fishermen from various parts of Kyushu also went to Korea, but the number of boats from Kyushu was far less than those of Hiroshima and Yamaguchi.[14]

Behind the rapid increase in the size of fishing fleets were direct subsidies from both local and national governments toward the purchase of boats, fishing gear, and processing equipment. In particular, Hiroshima Prefecture was ardent in its support of long-distance voyaging.[15] Together with the Fishery Bureau of the Ministry of Agriculture and Commerce and the part government- and part public-owned fisheries organizations (in a cooperative venture to promote fishing voyages to Korea), the Hiroshima prefectural offices provided necessary information and subsidies to those who wished to go to Korea. As a result, Hiroshima sent one of the largest fleets. In the meantime, the Japanese government, with strong intentions to expand Japan's sphere of influence among its Asian neighbors, also supported the advancement of Japanese fishing into Korea by creating favorable fishery laws. The Distant Water Fishery Promotion Act of 1897, for instance, regulated financial assistance as well as access to special programs for improving fishing boats and the skill levels of crews. A new law passed in 1900 exempted the Japanese going to Korea to fish from having to carry passports; in 1904, the Japanese government rescinded regulations preventing Japanese from entering certain Korean waters and made the entire sea area accessible to them. Because a law in 1908 allowed Japanese residents in Korea to engage in commercial fishing, some of those who had traveled to Korea from Japan began to move their base to Korea. Japan's annexation of Korea in 1910, in particular, expanded the scale of Japanese fishing and invited approximately 5,000 fishing vessels with more than 21,000 Japanese fishermen to Korean waters.[16]

The colonization of Korea was not, however, always advantageous for fishermen operating with limited capital because the integration of Korea into the orbit of rapidly expanding Japanese capitalism resulted in the development of capitalistic fisheries with large-scale equipment and many crew members. Small fishermen had to compete with these gigantic fisheries' capital. In addition, the massive entry of Korean fishermen had made the already fierce competition fiercer.

The increasingly crowded Korean waters, full of both large and small fishing vessels, had forced some fishermen to go to alternative sites. Among several new, promising destinations, Taiwan had bright prospects of expanding markets and a rapidly growing Japanese population after Japan's victory in the Sino-Japanese War (1894–1895) and the following cession of Taiwan to Japan. Fishermen from Okikamuro advanced into Taiwan, particularly Keelung and Kaohsiung, in the late 1890s. At first, they built a fishing vessel at home and sent it by steamboat. Later, they started building ships in Taiwan, where only a captain resided and crew members went to and from Okikamuro between late summer and spring. They engaged in pole-and-line fishing for sea bream as well as longline fishing for tuna, skipjack tuna, spearfish, and shark.[17] The colonial government of Japan in Taiwan played an aggressive role in luring Japanese fishermen by promising financial support, building facilities necessary for fishing, and improving the infrastructures of fishing ports and districts of residence.[18]

This indicates that the expansion of Japanese colonialism accompanied the enlargement of the Japanese fishing sphere, but its expansion was not always restricted to the extent of Japan's empire building. The people of Momojima and Tashima islands, which were located side by side off the coast of Fukuyama in Hiroshima Prefecture, went to Manila Bay, under US control since 1904, and conducted *utase* net fishing. With superior fishing skills and keen economic prowess, they instantly succeeded in business and established themselves in the local fishing industries. Unlike those who went to Korea with official support, they sought assistance primarily from wealthy patrons in their home villages. Americans in the colonial government of the Philippines tolerated Japanese fishing activities primarily because the United States trivialized the significance of the Philippines in their world politics, and the officials of the colonial government did not deem the Japanese fishermen a serious threat to the Philippines.[19]

The expansive energy of Japanese fishermen seemed unlimited and continued to extend the sphere of their activities within and outside the Japanese empire.

Japanese women's contributions to the fisheries

Discussing the advancement of Japanese "fishermen" into distant waters, with the assumption that fishing is a man's business, casts only a half-light on the fishing industry in general. As analysis of any fishing society in the world would show, women played active roles in catching fish. Almost all women, both young and old, in fishing households harvested various kinds of seaweed. They went into shallow waters, collected seaweed, spread it on the beach to dry, and packaged the end product for sale. In addition to gathering seaweed in shallow waters, some fisherwomen accompanied long-distance voyaging crews. Although a widespread faith among fishing communities in the goddess Funadama, who was believed to protect ships, discouraged women from boarding ships in order not to arouse jealousy in the goddess, the strength of the taboo against women's presence at sea varied according to the size of the boat and the region. In Kanezaki, in Fukuoka Prefecture, for instance, many women crewed fishing boats as *tomo-oshi* (pushing the stern) and helped their husbands with steering and other jobs. Large boats for snapper and squid fleets did not include women, but their presence on smaller ones was far from unusual. Once boats returned to port, the women found themselves busy removing fish tangled in nets and grading the catch according to size and quality. Dragging a beach seine and collecting seaweed were also women's jobs in Kanezaki. Together with men's jobs, women's contributions to fishing operations and handling fish were indispensable to make Kanezaki one of the most prosperous fishing communities in Fukuoka Prefecture.[20]

People on *ebune*—literally, a house ship—also put a higher priority on improving efficiency by working in couples than on their fear of provoking Funadama. These people lived on fishing boats equipped with *tatami* mattresses and led nomadic lives as family units. This fishing style existed only in limited areas of a calm sea, in particular, the Seto Inland Sea, and deep, landlocked bays in Kyushu.[21] The nomadic lifestyle of *ebune* people does not mean they simply wandered around on the water. Instead, most of the *ebune* boats had fixed destinations, with specific catches, and maintained relationships with their hometowns by coming back to reunite with people of the same district during the Bon and New Year periods;

usually, they married during these days. A woman of the *ebune* supported family life by rowing and fishing with her husband, conducting house chores, and taking care of the children. It was also her job to negotiate borrowing a bath and finding a place to wash. She stayed on the boat through pregnancy and childbirth. Although most *ebune* boats operated in calm inner seas or gulfs, those of Kanezaki in Fukuoka often plowed through the rough waves of Genkai-nada to Tsushima in the medieval period, before becoming sedentary in the seventeenth century. After the dawn of the Meiji era, the *ebune* of Hiroshima Prefecture went to the island of Cheju in Korea with the financial support of the prefectural government. Of course, women (and maybe children) participated in the expedition.[22]

Besides fisherwomen working together with their husbands, Japan's "sea women," called *ama,* had engaged in diving in the open sea to catch abalone, Ceylon moss, conchs, turban snails, agar seaweed, sea cucumber, and other marine products since ancient times. Some of their catch was deemed precious. In particular, abalones gathered by *ama* in Shima, in Mie Prefecture, were treated as holy and were offered in the prestigious Ise Shrine dedicated to Amaterasu, the sun goddess, and the imperial court. The importance of their role as food providers to holy entities reflects the unbreakable nexus between Shintoism, maritime culture, and women. The contributions of *ama* were significant enough to earn protection from both the Ise Shrine and the secular power of feudal lords in Ise and Shima.[23]

As fishermen and women daringly ventured to distant places, the *ama* divers had occasionally shown a similar disposition to travel away from home. Diving was a special skill requiring years of training, and these girls started diving from about from the age of ten, under their mothers' tutelage. Once they reached adulthood, with professional diving skills, these women left home, leaving the house and children in the hands of their husbands. Their journeys sometimes continued through the year, although those who had many children or old *ama* stayed home and dived from nearby shores.[24] The destinations of the *ama* included various sites in Japan, from the island of Rebun off the coast of Hokkaido to the coast of Tsushima. In particular, the *ama* of Shima and Ōura, in Yamaguchi Prefecture, went to the islands of Cheju and Ullung, and the southern coasts of the Korean peninsula. In 1900, for instance, forty fishing boats carried *ama* from Shima to Korea to gather Ceylon moss and abalones.[25] Kanezaki of Fukuoka Prefecture, with a reputation as a cradle of *ama* divers in Japan, spread throughout various parts of the Sea of Japan, including the island of Hegura, about thirty miles off the coast of Wajima city in the Noto Peninsula; the islands of Oki, Tsushima, and Iki; and the Gotō Archipelago.[26] Interestingly, as an ancestral land for *ama* descendants, Kanezaki still stimulates exchanges of personnel with these places; some of the women have even married Kanezaki men.[27]

While engaging in fishing and diving, *ama* women divers and *ebune* fisherwomen traded their own catch. Aquatic products were not staples but items used to barter for rice, vegetables, and other food and daily items. Their primitive bartering later evolved into cash sales. The money brought the women the authority of cash income within their households as indispensable contributors to their families' incomes. As a young *ama* on the island of Hegura stated, "We will never let our husbands have an affair, because we work very hard."[28] Some of the *ama* and *ebune* fisherwomen became full-time peddlers. In Tsuru, in Oita Prefecture, a wife who was a host for *ebune* boats made a voyage and went ashore to sell tangerines in spring, mackerel in early summer, and kitchen and table utensils in other seasons, while leaving her children in the care of her husband on a moored boat.[29]

As the negotiation and trading skills of women were essential to the survival of *ebune* society, their presence in trading was noteworthy in the fishing economies. Although only a limited number of women became *ama* or fished on *ebune* boats, female peddlers were omnipresent among fishing villages throughout the nation. Anthropologists Dona Lee Davis and Jane Nadel-Klein maintain that women in almost every fishing community in the world are heavily involved in fish processing and marketing.[30] Japanese women in fishing villages have played an essential role in processing and marketing the catch that men bring to shore. Instead of letting others meddle in their family businesses and losing money, these women preferred to handle their husbands' haul directly to make clear profits. In order to sell fish at higher prices, they worked to keep and sell fish as fresh as possible, and added extra value to the commodities by processing them into gutted, dried, pickled, salted, and baked products.

In the castle town of Hiroshima city, most fish peddlers were women from Nihojima and other fishing villages, and their domination of the fishing business remained unchanged even after the Meiji period; during the early Taisho era (1912–1926), 148 out of 168 traders were women.[31] When walking around to sell their wares, they called out *"nanmaē,"* and this voice and tone became well known in downtown Hiroshima. In Suō-

Ōshima, women called *katagi* peddled fish throughout the island. Many female peddlers put fish in baskets or tubs and carried them on their heads, while those in Susami of Wakayama shouldered their products.[32] These women occasionally diversified their commodities by adding vegetables and fruits, kitchen utensils, writing brushes and ink, kimono fabric, pottery, and various other products to their line of goods. Usually, they visited the houses of regular customers as well as selling to other people; they avoided stealing customers from each other. When a woman closed her business or retired, her daughter inherited her list of patrons. These women were, as folklorist Kiyoko Segawa called them, "carriers of Japanese culture," spreading marine products and other goods throughout the nation. Thanks to the women, fish have been deeply integrated into the dietary habits and religious ceremonies of the Japanese, even in the mountains.[33]

The women peddlers' markets sometimes extended beyond nearby towns into foreign countries. One of the most notable groups with an extensive sphere of commerce were the women called *otatasan*, in Masaki, Ehime Prefecture. Their sales routes covered the entire nation, and some of them reached the Kuril Islands, Taiwan, Korea, the northern part of China, and Manchuria. When the fishermen of Momojima and Tashima islands went to Manila, some wives joined them to support their businesses. The wives of fishing boat owners managed sales and took care of crew members, and those of ordinary fishermen peddled fish around Manila and coastal towns when their husbands were away in Manila Bay. Language barriers and other obstacles did not hinder their businesses, and their daily contact with the Filipino and Chinese customers went well.[34]

As the extent of their journeys indicates, they fulfilled their roles as messengers of Japanese maritime culture by weaving a trade network within and outside Japan. It is highly significant that women played major, if not always dominant, labor roles in fishing, processing, and marketing. Their engagement in income-producing jobs gave them a certain economic autonomy and authority within the household, while weakening the patriarchal control of male family members. Even women who led sedentary lives primarily took charge of the house and managed community matters, including local politics and the educational system, during the absence of men who did deep-sea fishing. Their contributions to the fishing industry and community service were noticeable enough to dismantle the notion of Japanese women as passive victims of a strict patriarchy.

As women's activities indicate, fishing communities developed different characteristics than those of agricultural areas. First, they were highly mobile and could relocate as fishing seasons and conditions dictated. Long-distance fishing enterprises often fostered the emigration of people from their home villages to bases nearer their fishing grounds, in order to reduce the cost of operating large vessels. In such cases, both men and women participated in forming branch villages, called *edamura*. By adapting well to their new environments through interaction with the local people, the men and women of the community established themselves in their specialties of fishing, processing, and peddling. Another distinctive characteristic of fishing societies was that they did not develop strict systems of primogeniture, as did agricultural communities. Due to the strong possibility of death at sea among fishermen, including heads of families, fishing households created a flexible inheritance system in which the youngest child would often inherit the family property.[35] Moreover, the fishing community was much more porous than the agricultural one, as indicated by the prevalent custom of adopting children and training them to be professional fishermen. Because fishing ports were not only departure points for people sailing out, but also stopping places for traders and fishermen coming from elsewhere, the community habitually interacted with strangers. The constant influx of fresh blood from outside supplemented absentee members at sea or on peddling journeys and constantly invigorated the community by bringing in new commodities, ideas, and information. The people of fishing communities, regardless of gender and place of origin, reinforced and supplemented one another and shaped the contours of social life.

Japanese fishermen spread throughout the Pacific

The fishermen in Wakayama Prefecture, who had developed the tradition of deep-sea fishing for centuries, increased the scale and speed of long-distance voyaging in many directions after the dawn of the modern era. The fishermen in Arida, Kaisō-gun, and Hidaka-gun, all of which were close to Osaka, went through the Seto Inland Sea to reach Korea, while those in Higashi-muro-gun went east through the Sea of Japan, the Kuril Islands, and as far as the Bering Sea to catch salmon and fur seals. In addition to these east- and westbound voyages, the Wakayama fishermen opened a new route of travel across the vast Pacific Ocean: the passage to Australia. Usually, fishing boats forged into open waters by following islands; the route to Korea by way of the islands of Iki and Tsushima typifies this pattern. Yet, as anthropologist Akira Gotō states, fishermen occasionally crossed a vast ocean without stopping along the way and reached remarkably distant places, where natural conditions were completely different from those of their home waters.[36] Whether or not the destinations were within reach of Japanese imperial control, fishermen traveled to find better fishing grounds without official permission.

It was precisely in such cultural circumstances that vast numbers of Japanese fishermen went to Australia and engaged in diving for pearl oysters beginning in the late nineteenth century. In the late 1870s, a Japanese man named Kojirō Nonami (aka Nona) from Shimane Prefecture achieved fame as an outstanding diver on Thursday Island.[37] Thursday Island, a small island off the coast of Darwin in the state of Queensland, faced the Arafura Sea and was rich in pearl oysters. In those days, pearl oysters were valued as raw material for high-quality buttons; there was a great demand for the buttons in Europe. In the early twentieth century, Australia produced almost 90 percent of the pearl oysters in the world, and skillful divers were always in short supply to meet the increasing demand. Entrepreneurs targeted and began to aggressively recruit Japanese divers. During the early 1890s, hundreds of Japanese men began traveling to Thursday Island, most of them from the coastal communities of south Wakayama, specifically areas such as Tanami, Susami, Nachikatsuura, and Shingū.

The connection between Wakayama and Thursday Island began when British engineers came to Kashino-zaki and Shio-no-misaki, at the southern end of Wakayama, to construct beacons in 1869. After that, they took several local young men to Kobe, where foreigners lived. The news that remunerative work was available in Kobe inspired the young men of the town to dream beyond their current reality, get hired by a foreign company, and eventually go to Australia. As early as 1894, the island had a Japanese club with 346 members, of whom 254 (72 percent) were from Wakayama, 22 (6.4 percent) from Nagasaki, 15 (4.3 percent) from Hiroshima, and 8 (2.3 percent) from Fukuoka prefectures. The number of Japanese divers increased year after year, despite the extremely dangerous diving conditions, as indicated by the high casualty rate: the death rate among Japanese divers reached 10 percent between 1908 and 1912. Nevertheless, it was a highly lucrative job compared to those available at home, and the divers' determination to succeed financially and their insatiable drive to dominate nature surpassed their fear of death.[38] In 1897, the number of Japanese divers rose to more than a thousand, and Japanese-owned pearl-fishing boats accounted for more than 15 percent (about thirty boats) of all the boats around Thursday Island.[39]

Some of the Japanese who left for Thursday Island moved up the social ladder by running their own businesses. Torajiō Satō was the most notable figure. Born in Motoizumi, Musashi (now Saitama Prefecture), in 1864, he was educated at the University of Michigan and settled in Wakayama after marrying the daughter of a local timber dealer there. When he went to Australia in 1893, at the request of Japanese foreign minister Munemitsu Mutsu, to investigate the conditions of Japanese immigrants, he saw the great potential of the pearl oyster business in the Arafura Sea. Soon after, he moved to Thursday Island with several young men from Wakayama and started a pearl oyster business. His new enterprise prospered. Nicknamed the King of Thursday Island, Satō owned dozens of boats and hired more than 1,800 employees. However, as the Japanese presence in the industry increased, anti-Japanese sentiment intensified among their white rivals. Soon, Japanese exclusion took shape in a law promulgated by the state of Queensland, according to which only naturalized British citizens were allowed to possess diving boats and conduct boat-lease businesses. In 1900, Australia set a strict limit on nonwhite immigration in order to strengthen its white Australia policy. This policy devastated Satō's business and forced him to go back to Japan. But the legal maneuver failed to completely expel all Japanese divers from Australian waters. Afraid of losing an excellent workforce, the white boat owners' association lobbied the government to allow a certain number of Japanese immigrants. Furthermore, some Japanese illegally smuggled themselves onto Thursday Island and Bloom, on the east coast, by way of Shanghai, Hong Kong, and Indonesia, and engaged in diving in the powerful and deadly current.

These reckless men, however, did not dare to bring their wives to Australia, because the exclusion policy allowed only wives of government officials, travelers, doctors, and wealthy merchants to come legitimately. Naoichi Hama of Kushimoto in Wakayama reflected on his days in Australia:

> Many young men in Kushimoto went to Australia. When they came back with substantial sums of money, they built houses and got married. After spending all their savings, they went back to Australia and worked again. Wives lived as if they were widowed. . . . I could never take my wife with me. Even I, as a man, had a very hard time landing in Australia, how could

she do so? First of all, no ship would let women on-board, except those with a powerful patron or under special circumstances. What's worse, there were no decent jobs available for women there. Therefore, we left wives behind. Once we arrived in Australia, we could not go home for at least two or three years. Quite a few of us never went home. I guess many men from Shionomisaki and Kushimoto ended their lives in Australia without seeing their wives again.[40]

His remarks reveal that pearl oyster diving in Australia was extremely lucrative and a couple of years' work brought the divers enough money to build a house, get married, and raise a family. Australia's positive reputation among the people of south Wakayama as a profitable land enticed many young men. However, Hama's memory also shows the negative by-products of working there, including the possible breakdown of family life. Because of the strict Japanese exclusion policy and scarcity of local job opportunities for women, Australia remained a place of *otoko-kasegi,* where only men went for work. In the meantime, the wives stayed in Japan and waited for their husbands for a long time, sometimes in vain.

When Australia turned a hostile eye to the Japanese divers, Hawai'i started drawing people's attention as a new and promising fishing ground. After Hawai'i introduced government-contract labor importation in 1885, it had primarily attracted farmers from Hiroshima, Yamaguchi, and other prefectures, most of them in western Japan. Throughout the contract immigration era, which ended with the overthrow of the Hawaiian Kingdom in 1893, the Japanese immigrants were contracted to work in sugarcane fields. In contrast to the great number of farmers who responded to the recruitment from Hawai'i, few full-time fishermen in Hiroshima and Yamaguchi exhibited any interest in going to Hawai'i. They were completely occupied with fishing around their home waters or the seas of neighboring nations, such as Korea, Taiwan, and the Philippines. However, Gorokichi Nakasuji, a fisherman from Tanami, in Wakayama, looked at Hawai'i in a different way. Born to a family that had fished for generations, he grew up to be a professional fisherman. When his grandmother died, he inherited family property from her. Like many other young men in south Wakayama, he also dreamed of going to Australia and making a fortune. But the Japanese exclusion policy dashed his original plan and compelled him to think about an alternative.

It was a letter that diverted his attention to Hawai'i, in the middle of the Pacific. After Gihei Kuno, of Mio village in Wakayama, went to Canada in 1885 and discovered that salmon were swarming in the rivers, people from the same and neighboring villages followed him. Among them was Nakasuji's friend Mankichi Murakami. When he stopped in Hawai'i on his way to Canada, he saw that Native Hawaiians were engaging in "primitive" skipjack tuna fishing and gave a detailed account of it in a letter to Nakasuji. This firsthand description of fishing in Hawai'i caused Nakasuji to recognize its huge potential for skipjack tuna fishing. He built a fishing boat 31.8 feet long by 5.8 feet wide, and, with a variety of fishing equipment, loaded his new boat onto a steamship bound for Hawai'i. He boarded a separate ship, the *Nan'yōmaru*, with his wife, child, and two fellow fishermen, and sailed for Hawai'i in 1899.[41]

2 Japanese Fishermen Enter Hawaiian Waters

The Formative Years of Commercial Fishing in Hawai'i and the Rise of the Japanese, from 1899 to the Early 1920s

Japanese fishermen arrive in Hawai'i

When Gorokichi Nakasuji arrived in Honolulu Harbor in December 1899, Honolulu was in great turmoil following an outbreak of the plague and its aftermath. A fire set to burn a small, contaminated section of Chinatown soon went out of control and wiped out the whole town. Thousands of people, including many Japanese, were left homeless. The tragic fire delayed the process of immigration and Nakasuji was detained for three weeks. Once he was released from the immigration office, a scarcity of residences due to the great fire forced him to move from place to place, without enough time to start his fishing business. During the unexpected three-month wait, he saw *nehu* (Hawaiian anchovy, *Stolephorus purpureus*) and *'iao* (silverside, *Pranesus insularum*), baitfish for *aku* or skipjack tuna (*Katsuwonus pelamis*), swimming in large schools in the harbor, confirming his belief in the great potential of Hawaiian waters.

Despite the bright prospects for a large-scale fishing enterprise, there were only four or five wooden boats operating within Honolulu Harbor at the time. Hawai'i did, however, have previous experience in commercial fishing. During the first half of the nineteenth century, the whaling industry flourished, and about 17,000 ships arrived in Hawai'i from the United States and Great Britain.[1] With the advent of petroleum to replace whale oil, the whaling industry died out and only traditional native fishing remained. In 1900, the total annual landing of skipjack tuna, for instance, was only 190 tons.[2] Among the small fishing population, Japanese fishermen had made their modest presence known since the onset of government-contracted immigration. On February 8, 1885, the steamer *City of Tokio* arrived at Honolulu Harbor with more than 940 Japanese passengers, many from Hiroshima and Yamaguchi prefectures, who were soon to become an integral part of the labor force to plant, harvest, and cultivate sugar. Most of the immigrants were from farming families, but these prefectures, especially the coastal areas along the Gulf of Hiroshima and Suō-Ōshima, in Yamaguchi—both of which had distinguished themselves by sending large numbers of migrants to Hawai'i—included many shore communities where people engaged in both farming and fishing. It was, therefore, natural that some of the immigrants with previous fishing experience took to the sea when they had time off from plantation work, and switched their full-time occupations to fishing after the expiration of their contracts. Umatarō Nakamura, one of the passengers on the *City of Tokio*, from the island of Heigun near Suō-Ōshima, was one of the Japanese who chose fishing instead of staying with plantation work. After working on a plantation on the island of Hawai'i for three years, he went to Kaua'i and built a fishing boat. "Because I was confident in fishing," he said, he started fishing and immediately made profits. "It was never uncommon to catch $7 to $8 worth of fish in 30 minutes," he said.[3]

The majority of the Japanese immigrants during the early days were single men, but women also engaged in fishing and peddling. Nobu Kurihara, from the island of Kasasa, off the coast of Suō-Ōshima, came to Hawai'i on the *City of Tokio* with her husband and children. After working at a sugar mill on the island of Hawai'i, her family moved to Honolulu in order to provide their children with a better educational environment. Disappointed that no good jobs were available in the capital city of Hawai'i, the Kuriharas started fishing in northeast O'ahu and farming on a leased ranch.[4] While Nobu Kurihara contributed to her family's income working as a fisherwoman, Jinkurō Ōhara's wife helped her family's finances by peddling. When Jinkurō, from Agenoshō of Suō-Ōshima, sailed out and filled his boat with fish, his wife sold the catch on plantations on the island of Kaua'i and quickly made profit.[5] These stories suggest that early Japanese immigrants brought with them from their mother country gender roles distinctive to coastal fishing areas, where men as well as women cooperatively fished and peddled before the advent of modern commercial fishing and distribution enterprises.

Hawai'i, in those days, also had Chinese fishermen, who primarily engaged in net fishing in shallow water. They were not in serious competition with the Japanese because they primarily caught their favorite fish,

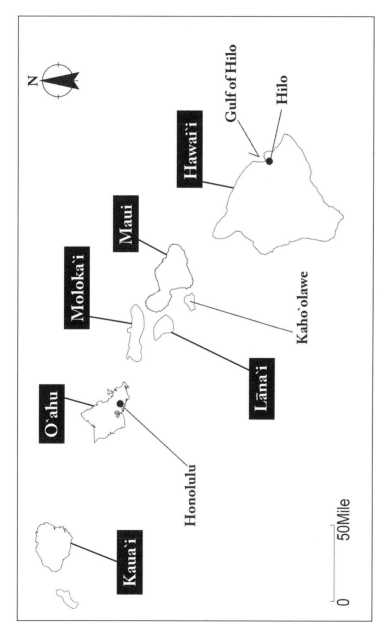

Hawaiian Islands.

mullet, in which the Japanese were not greatly interested. Their net fishing, however, came to an end when a ship cut a net and caused injuries. After that tragic incident, the Chinese shifted from fishing to managing mullet fishponds they obtained from Native Hawaiians.[6] The Hawaiian culture had developed ponds for storing and fattening mullet and milkfish, and large fishponds served as status symbols for chiefs, who could command many workers to construct and maintain them.[7] During the monarchical period, pond fishing remained an important industry in Hawai'i, and the pond mullet was highly valued. As of 1900, there were still 103 ponds from which fish were taken for commercial purposes, and the Chinese gained entry into this business by buying ponds from natives.[8]

In contrast to inshore fishing, in which the Japanese and Chinese played a part, offshore fishing was completely dominated by Native Hawaiians. Native Hawaiian fishing can be traced back to ancient times, when Polynesian mariners explored Oceania and eventually reached Hawai'i. They were able to cross the vast Pacific Ocean in part because of their highly skillful fishing. Polynesians traveled through the various archipelagos while consuming and trading fish for essential items, besides engaging in horticulture.[9] The acquisition of protein from fish and other marine resources was an essential aspect of early subsistence in the Hawaiian Islands, especially among the commoners, who consumed smaller amounts of pig, dog, chicken, and wild birds than the chiefs. Hawaiian fishermen set out to sea with single or double canoes accommodating two to six persons and engaged in pole-and-line fishing for skipjack tuna with the lure hook called *pā*.[10] Canoes were fitted with the *malau*, a bait holder containing live *'iao*.[11] Since they deemed the sea to be a great reservoir of food, they had a history of fishing traditions rife with various taboos, called *kapu*, on catching or eating certain kinds of fish. In particular, skipjack tuna was revered because, according to legend, it had saved an early voyager coming to Hawai'i from Tahiti from storms at sea by quieting the waters. Therefore, skipjack tuna fishing was under certain taboos, and its season was limited to six months a year. In fact, the actual reason behind the taboo was probably to protect spawning and juvenile seasons.[12]

Gorokichi Nakasuji deemed the off seasons imposed on skipjack tuna fishing to be nothing but a manifestation of native laziness, rather than a means of conservation. When he checked the water temperatures throughout the year, he was convinced that year-round skipjack tuna fishing would be possible. Because the fishing boat he brought from Wakayama could accommodate much larger amounts of baitfish than the Hawaiian canoes, and it was as efficient as seven canoes, his nonstop fishing greatly expanded the skipjack tuna supply and inevitably lowered prices to 25 percent of the original market value. The drastic fall in prices forced many Hawaiian fishermen to give up their businesses, and even drove some of them to make an attempt on Nakasuji's life while he was at sea. However, the plan was disclosed to Nakasuji by Native Hawaiian friends he had met through fishing. When native canoes started chasing him at sea, at the signal of the raising of an oar, he filled his sail and made his escape at full speed. His fishing boat was structured to sail faster than native canoes. The fishing grounds off southern Wakayama were about twenty-four to fifty miles from the coast. Therefore, boats were slim and lightweight, able to reach even the farthest fishing spots within three hours.[13] This swift boat that Nakasuji introduced to Hawai'i saved his life. "During the pioneering days, I had to stake my life on everything," said Nakasuji of the early years.[14]

This episode of near violence was a manifestation of the friction caused by the initial clash of different fishing cultures. Traditionally, Native Hawaiians had caught fish principally for the purpose of self-sufficiency, and the fair distribution of the catch had always been a concern of the community. The sharing took place not just within a village, but among the members of an extended family of communities, or *'ohana*, some of whom lived on the coast and some in the uplands. This interdependence of the land and the sea, embodying the ideal health and integrity of the *'ohana*, is often represented in traditional Hawaiian stories.[15] But this self-sustaining ideal was not compatible with the more exploitative fishing style the Japanese suddenly introduced into the seascapes of Hawai'i.

The other islands saw violent encounters between existing fishing communities and Japanese newcomers. Kanzaburō Ageno remembers that his father was excluded by Hawaiian and Chinese fishermen in his pioneering days on Maui. Young Ageno came to Hawai'i in 1908, together with his family. His father was a fisherman trained at a fisheries school in Osaka and engaged in skipjack tuna fishing in southern Kagoshima. No sooner had he started operations in Kahului, Maui, than he encountered various restrictions placed on Japanese newcomers, such as being banned from using certain ports even during storms, and limitations on operating in rich fishing grounds. Violators were subject to the confiscation and even destruction of their fishing boats by old timers. Ageno's father was experienced in sumo and judo, and he defended himself effectively when he was assaulted by those who attempted to throw him out.[16] Overcoming the friction with their counterparts of other ethnicities, Japanese fishermen gradually consolidated their position in the islands' fisheries, and provided aquatic resources to the increasing number of Japanese immigrants. In 1900, there were 50,000 Japanese in Hawai'i, nearly 40 percent of the total population. This number increased to 65,000 within nine years.[17] Many of the Japanese chose to follow their customary diet of vegetables and fish, and Hawai'i needed more aggressive methods of fishing in order to meet the increasing demand.

It would, however, be misleading to insinuate that Japanese fishermen had nothing to learn from the Native Hawaiians and that the interaction between the two groups produced only discord and animosity. In fact, they frequently exchanged aquatic knowledge, fishing methods, techniques, and gear. The Japanese newcomers enriched the local fauna by bringing previously unknown marine species to the market.[18] They also coined their own names for fishes in Hawai'i, such as *menpachi* for squirrel fish (*u'u* in Hawaiian, *Myripristis berndti* or *Argyromus*), *shibi* for bigeye tuna (*'ahi, Thunnus obesus*), and *aji* (*akule, Trachurops crumenophtahlmus*). Some of these names were integrated into the local glossary of fisheries. Around 1890, the Japanese also introduced throw-net fishing to Hawai'i, a method that was quickly adopted by the Hawaiians and used where waves broke on shallow reefs.[19]

The Hawaiians reciprocated this new knowledge and technique by showing the Japanese their fishing hook, called the *kenken*. The *kenken* hook, with a bird feather and a pearl oyster, was widely used along Polynesian coasts. Fishermen from Wakayama took this hook home and copied it, using pearl oysters brought from the Arafura Sea in Australia. They also learned the structure of a yacht and applied its stabilizing system to a fishing boat by installing a long board at the bottom of a boat and equipping it with three sails, in order to catch winds from any direction. Use of this boat, named the *kenken-sen* (*sen* meaning boat), swiftly spread around southern Wakayama and beyond, and was used in fishing operations until plastic boats replaced it in the 1960s. The *kenken* fishing hook, the collaborative product of the Hawaiian and Oceania fishing cultures, was used until recently.[20]

Gorokichi Nakasuji never stopped pursuing technical innovations and devising new fishing methods suitable to Hawaiian waters. For example, he pioneered the use of an electric lightbulb to lure baitfish. He also successfully worked out longline fisheries for *'ahi*, three species of large (more than sixty pounds) tuna (bigeye tuna, *Thunnus obesus;* bluefin tuna, *Thun-*

nus thynnus; yellowfin tuna, *Thunnus albacares*), after experiencing serious damage to his fishing equipment and losing his catch to sharks.[21] In Hawai'i, he was the first to install a gasoline engine to propel a boat. This event was truly epoch making, although it took a while before the majority of fishing boats in Hawai'i were gasoline powered. While developing innovative fishing techniques and gear, Nakasuji aggressively recruited plantation workers with fishing experience. Fishermen in those days earned $30 to $40 per week, while plantation workers made only $18, and it was easy to persuade them to join the fishing trade.[22] Coincidentally, the Organic Act, transferring the sovereignty of the Hawaiian Islands to the United States, terminated the contract system in 1900 and freed Japanese plantation workers to pursue any profession. Before this act, Gorokichi Yabe, who was born in Tanami in south Wakayama and came on contract to Hawai'i in 1895, immediately escaped from a plantation on the island of Hawai'i because his "true purpose to come here [Hawai'i] was fishing," he said, and he could not wait what might have been years for the law to change. He changed his name so he would not be dragged back to the plantation, and started skipjack tuna fishing off Coconut Island in Hilo.[23] Ironically, with the end of the contract system, such risky actions turned out to be unnecessary.

More than one transpacific highway

In addition to recruiting fishermen from the cane fields, Nakasuji encouraged approximately 250 people from Tanabe, his hometown, to move to Hawai'i, aggressively utilizing the nexus of his relatives and localities.[24] Inspired by the success of Nakasuji and the bright promise of the fishing industry in Hawai'i, fishermen from Wakayama started coming to Hawai'i one after another. In 1900, for instance, 259 Japanese were engaged in fishing in O'ahu, and by 1903, this number had increased to 707. During the same period, the number of Native Hawaiians at sea dropped from 654 to 533.[25] The rise of the Japanese and the decline of native fishermen occurred on other islands in the Hawaiian chain as well, especially on the Island of Hawai'i, Lānai, and Maui, where Japanese competition drove many native fishermen out of business.[26]

The route from Wakayama to the islands was far more complicated and multidirectional for fishermen than for farmers, who usually came directly from Japan. Due to Japanese nautical mastery and rich experience in navigating the ocean, more than one transpacific highway was available.

Actually, their movements could involve return, zigzag, circular, multiannual, seasonal, or permanent paths, as indicated by the story of Heikichi Komine. Komine was born in Kushimoto, in the southern part of Wakayama, and went to Sandakan, in Borneo, where he could land without a passport. In Sandakan, he obtained permission to go to the Philippines. During his three-year stay there, helping with his brother's business, he harbored an ambition to become a seaman. Soon he boarded a vessel bound for New York as a second engineer. He stopped over in Honolulu, where his cousin, who was living in Hawai'i, persuaded Komine to jump ship and join him. Komine dove into the sea at midnight. His cousin picked him up in a small boat and took him to Honolulu Harbor. Komine started working on a skipjack tuna fishing boat. "The law in those days was not very strict," Komine said. When customs officers caught him more than five years later, they pardoned him because he had already been in Hawai'i for years, had gotten married, and had children.[27]

His case was the tip of an iceberg of many fishing immigrants without valid papers. According to the recollections of Banji Hamaguchi, a fisherman from Susami, in Wakayama, who also jumped ship somewhere in the United States and engaged in fishing in San Pedro, near Los Angeles, many fishermen from south Wakayama came illegally to the United States. Some of them crossed the border from Mexico on foot, or sneaked into the United States on transpacific steamboats jumping off in Honolulu, Tacoma, Washington, or other ports.[28] These attempts were risky, said Kumao Ichimatsu, who was living in Seattle. He wrote a letter to a local newspaper in southern Wakayama and informed its readers of the dangers of jumping into the deadly cold sea around Seattle. He tried to discourage lower-ranking seamen from deserting ship in the United States because they would come under the strict oversight of immigration officials. He was from Nishimuro-gun, in Wakayama, and it wrenched his heart when he heard of men from his hometown jumping into the sea and freezing to death. Ichimatsu dissuaded the people of south Wakayama "with foolishly bold passion" from violating the law and risking their lives.[29] His warning implied that coming to the United States as a seaman and jumping ship during the night was a common idea among men from the coastal districts of south Wakayama, although the success rate varied according to the water conditions and the level of surveillance by immigration officers. Heikichi Komine was lucky because he landed in Hawai'i, where the water temperature was mild and immigration officials were more relaxed and tolerant than those in Seattle. Once in Honolulu, the integral network of Wakayama communities provided him with various means of support and enabled him to find a job and raise his family.[30] It was also common among men in southern Wakayama to transfer their permanent domiciles to Kanbara-gun, in Niigata Prefecture, which repeatedly suffered from flooding and had a larger quota of emigration, then legally make their way to Hawai'i.[31]

Because many fishermen from Wakayama planned to go to the mainland and Canada, where the Japanese were making steady headway in local fishing enterprises, Hawai'i served as a primary stepping-stone on their transpacific route. Nevertheless, their presence in Hawai'i resulted in the formation of a sea squadron in Honolulu called Kishū-katsuosen-gumi (Kishū Skipjack Tuna Fishing Fleet). As of 1907, the Kishū fleet consisted of eight boats that operated far off the island of Moloka'i in the summer and fall and off the 'Ewa coast of O'ahu in the winter. Their aggressive fishing brought enormous amounts of skipjack tuna to the local market, and the value dropped to 10–20 cents per whole fish, changing it from a rare, expensive fish into one of the most affordable.[32]

The rise of fishermen from Yamaguchi and Hiroshima prefectures and the birth of Japanese fishing companies

As fishermen from Wakayama Prefecture worked their way into skipjack tuna fishing, those from Yamaguchi and Hiroshima prefectures gradually rose to prominence in other styles of fishing, mostly the longline method and net fishing to catch bottom fish. In particular, fishermen from Okikamuro, in Yamaguchi Prefecture, enhanced their status within the Hawaiian fishing community. The fishermen from Okikamuro, a small island with a population of 1,000 households at most, however, did not advance to Hawai'i with the beginning of Japanese immigration in 1885 because many of them, in the last two decades of the nineteenth century, were busy migrating to various parts of the Seto Inland Sea, as well as to the coasts of Kyushu and Korea. In particular, the fishing fleet to Korea, called Chōsen-gumi (Korea group) played a leading part in all the long-distance fishing ventures from Okikamuro in those days. But its heyday of westbound journeys abruptly ended around 1902. The immediate cause was the shipwreck of two vessels and the death of their crew members. The indirect, yet far more devastating, reason was overcrowding in Korean waters. As of 1900, the Okikamuro fleet made up only 3 percent of the 2,000 to 3,000 Japanese fishing boats that flocked to Korea; with the passage of time, the Okikamuro fleet dwindled further. Overwhelmed by the huge number of their rivals, the fishermen of Okikamuro inevitably headed to Taiwan, Chingtao in China, and, finally, Hawai'i during the first decade of the twentieth century.[33]

Once landing in Hawai'i, the Okikamuro fleet found itself amid a wealth of fish and invited fellow fishermen from their home island to join them. In 1908, those from Okikamuro and its neighbor, Agenoshō, in Suō-Ōshima, stood out among the fishermen operating in the Honolulu area. Their lifestyle was, however, far from temperate. According to one fisherman from Okikamuro, "They drink before fishing, and, if the catch is poor, they drink again to change their luck. Of course, they drink a lot to celebrate rich hauls. Since they went on sprees with geisha girls, they spent all the money they had."[34] He believed they wasted their money on alcohol due to the peculiar social arrangement of their bachelor community. "If they had wives and children, they would be different," he said.[35]

The frequent socializing and drinking with other fishermen was financially disastrous and undermined their health, but was also instrumental in strengthening and stimulating horizontal transfers of knowledge among their members. The solid network of the Okikamuro fleet, nurtured through frequent socialization, allowed its members to share indispensable navigational information, such as sea currents and weather conditions, in order to explore new fishing grounds. Thanks to the collective knowledge they accumulated and shared, they successfully exploited foreign waters and rapidly expanded their enterprises. Around 1915, there were 46 men from Okikamuro in Honolulu, 57 in Hilo, 4 on Maui, and 5 on Kaua'i; within four years, the number expanded into 84 in Honolulu, 90 in Hilo, 13 on Maui, and 9 on Kaua'i. Okikamuro sent many islanders abroad, but 196 men in Hawai'i far surpassed the 110 in Taiwan and 95 in Korea.[36] Many of them were accompanied by their wives, children, parents, and siblings, and the total population of the community became much larger. The gradual influx of family members into Hawai'i reduced the problems inherent in a bachelor society.

Among the early immigrants from Okikamuro, Kamezō Matsuno and Isojirō Kitagawa were the most prominent. Matsuno came to Hawai'i in 1902, at the invitation of his uncle, a fisherman operating out of Hilo. Matsuno's personal record in Hawai'i does not include plantation work. He first worked at a liquor shop for a white owner while attending a boarding school, and later began peddling fish. He worked hard, waking up at 4:00 a.m. and working until 11:00 p.m., accumulating enough capital to venture into a larger business. Unlike Matsuno, Kitagawa worked at a plantation on the island of Hawai'i. After going back to Japan to serve in the Japanese army during the Russo-Japanese War, he returned to Hawai'i and started selling fish in Hilo.[37] A small fishing camp faced the Gulf of Hilo, on the island of Hawai'i, and it gradually developed into one of the major fishing stations on the island with a number of fishermen from Okikamuro. As early as 1901, they established a rotating credit association called *kō*.[38] A *kō*, often called *tanomoshi*, was an association of people who made regular contributions to a common economic pot, which primarily provided the capital necessary for the startup or expansion of small businesses, major purchases, or recreational activities because few banks were available in the early 1900s. Even if a financial institution was available, it was difficult for Japanese immigrants to obtain loans.

The monetary support from such a mutual aid association and the close-knit relationship among the people of Okikamuro enabled Kamezō Matsuno and Isojirō Kitagawa to strengthen and broaden their businesses. They created Suisan Co., or Suisan Kabushiki Kaisha, usually called Hilo (Waiakea) Suisan, in 1907, with $1,250 in capital.[39] This was the first Japanese organization for buying, processing, and marketing fresh fish in Hawai'i.[40] Kikumatsu Kadota, from Okikamuro, was its first president, and Matsuno and Kitagawa held executive positions. About fifty fishermen and fish peddlers, many of whom were from Okikamuro, bought $5 shares to support the new company. Those from Hiroshima Prefecture participated in management as well as owning stock in the company. The second company president, Heitarō Egawa, for example, was from Nihojima, in Hiroshima city, a village where people engaged in fishing and farming until they lost nearby fishing grounds due to massive landfills in the Gulf of Hiroshima.[41] Although some fishermen from Nihojima traveled to Korean waters, many found a new way out of their difficulties, immigrating to Hawai'i instead of competing for diminishing resources in their crowded home shores or Korean waters.[42]

In Honolulu, the first Japanese fishing company came into existence in 1908. Unlike Hilo, where fishermen and fish dealers created the Hilo Suisan, Japanese journalists and other community leaders played a major part in the formation of a new fisheries organization in Honolulu. According to Yasutarō Sōga, owner of a Japanese newspaper company, Nippu Jijisha, the Chinese dealers benefited most from the business, despite the

Fishing boats on Waiakea River, Hilo, island of Hawai'i. Photo courtesy of Hawai'i State Archives.

Japanese monopoly on the fishing enterprises. The Japanese fishermen grumbled at their possible profit taking but remained powerless against the Chinese monopoly in the distribution of fresh fish.[43]

As Sōga stated, the Chinese dominated fresh fish sales in Honolulu. A Chinese businessman, Chung Kun Ai, constructed the first privately owned fish market in Honolulu in 1904. Before its appearance, the Hawaiian government owned and managed a marketplace that seems to have been a "large, well-appointed, and well-administered institution."[44] However, complaints from Chinese fish dealers that the government fish inspector was cheating them spurred Ai to create a rival institution called the City Market. The following year, Anin Young, a Chinese entrepreneur, and several other influential Chinese established the O'ahu Fish Market alongside the City Market, but nearer King Street and the tram car service. The better location of the new market attracted most of the fish vendors, which resulted in the closing of both the government market and the City Market.[45] Thereafter, the Chinese merchants in the O'ahu Fish Market, the only remaining marketplace in Honolulu, dealt exclusively with the fishermen's catch and took 10 percent of the sales as commissions. Yasutarō Sōga regretted that the Japanese fishermen submitted meekly to Chinese control and, worse, wasted their money on drinking. Sometarō Shiba of Hawai Shinpōsha, another Japanese newspaper company; Matsutarō Yamashiro of Yamashiro Hotel; and Toshiyuki Mitamura, a medical doctor who came to Hawai'i in 1889 from Wakayama at the invitation of the Hawaiian government, supported Sōga and started a campaign to educate Japanese fishermen to improve their lifestyle and convince them to create an organization representing their own interests. Sōga's caravan frequented Kaka'ako, a fisherman's town facing the Kewalo Basin, the central home port for commercial fishing boats, and addressed Japanese fishermen and their families about the importance of seizing power from the hands of the Chinese.[46]

Matsutarō Yamashiro, who was part of the enlightenment campaign, came to Hawai'i in 1890 from Nihojima as a contract immigrant and worked at two plantations for six years. With the savings from his hard work and extremely frugal life, he opened an inn, the Geishin-ya Ryokan, on Nu'uanu and Kukui streets in Honolulu in 1896. It was, however, reduced to ashes in the Chinatown Fire of 1900. Four years later, he started the Yamashiro Hotel on Beretania Street and College Walk. During the 1909 Japanese plantation workers' strike on O'ahu, he served as treasurer of the Higher Wage Association and offered his hotel as the site for meetings.[47] As his involvement in the association indicates, Yamashiro had a deep sense of public duty and cultivated connections with many influential Japanese on O'ahu. Soon he expanded his concerns beyond the problems of land industries to those of the sea and fisheries and supported Sōga's efforts.

The *Tatsumaru* incident in 1908 made the creation of a Japanese-owned company a pressing concern. The seizing of the Japanese steamship *Tatsumaru* by a Chinese patrol boat off the coast of Macau triggered a strong protest from the Japanese government, even though the ship had attempted to smuggle firearms and ammunition. Succumbing to pressure from Japan, the Chinese government formally apologized and released the *Tatsumaru*. This compromise so angered the Chinese people that their boycott of Japanese products spread to Hawai'i, where Chinese merchants refused to buy fresh fish from the Japanese.

The urgency of securing an outlet for the Japanese fishermen, coupled with a strong push by community leaders, produced the Hawaii Fishing Co., with $50,000 in capital, in September 1908. The twenty-one

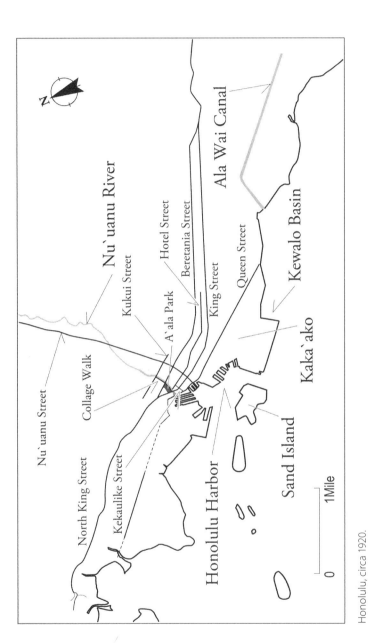

Honolulu, circa 1920.

Matsutarō Yamashiro. Photo courtesy of Eloise Yamashiro Kurata.

founding members included Yasutarō Sōga, Sometarō Shiba, and Toshiyuki Mitamura. Although Mitamura did not have any fishing background, his strong entrepreneurial sense compelled him to participate in the startup of the company and to become its the first president. The new company opened the King Fish Market on Kekaulike Street for wholesaling and retailing.[48]

Two years later, Matsutarō Yamashiro established the Pacific Fishing Co., with $10,000 in capital.[49] The striking difference between the two was the degree of Japanese managerial control. Unlike the Japanese dominance of executive positions at the Hawaii Fishing Co., Yamashiro's company included Chinese in managerial posts. The collaborative management kindled anti-Chinese sentiment among Japanese fishermen and discouraged many of them from bringing in their catch. Through considerable effort and service to the fishermen, such as buying excess harvests of small fish, with which the Hawaii Fishing Co. was unwilling to deal, Yamashiro gradually won the trust of Japanese fishermen and expanded his enterprise.[50] In 1914, another Japanese fishing company, the Honolulu Fishing Co., was born, with $5,000 capital; Chōzaemon Nakafuji of Yamaguchi, Tsurumatsu Kida of Esunokawa, a small town in south Wakayama, and other Japanese served as intermediaries to assist in its formation.[51] Kida, a former apprentice of Gorokichi Nakasuji, excelled in skipjack tuna fishing. While owning and operating two fishing boats—the *Kasuga-maru I* and the *Kasuga-maru II*, named after Kasuga Shrine in his hometown—he was also aggressively involved in community activities, including organizing a mutual support association for those coming from Esunokawa, creating a fishing boat owners' organization, and establishing the Wakayama-ken (*ken* meaning prefecture) association. Supporting undocumented immigrants was also one of his main activities.[52] As Kida greatly concerned himself with the interests of Wakayama, the Honolulu Fishing Co. had larger numbers of affiliated fishing boats owned by fishermen from Wakayama than the other two companies.[53]

The establishment of fishing companies and the rivalries among them stimulated the development of the Japanese fishing industry in Hawai'i remarkably, since these organizations functioned as hubs connecting fishermen and dealers and as the center of the fish distribution system. The companies charged 5–10 percent commission on the haul of the boats before any other deductions, such as fuel, bait, and food, were made.[54] Besides auctioning off the landed hauls and securing the benefits, the fishing companies functioned as patrons of the fishermen. They financed necessary expenses for fishing operations, including ice, fuel, and food. Accommodating or serving as guarantors for bank loans to build boats was also one of the important jobs of the companies. Thanks to their financial support, many fishermen possessed their own fishing boats. Kewalo Basin was soon filled with wooden Japanese-style boats called sampans.[55] The companies paid the registration fees and wharf fees for each new fishing boat. In return for this assistance and support, the fishermen were obliged to sell their haul to their sponsors.[56]

The supportive characteristics of the fishing companies resembled those of fish merchants in Hiroshima and Yamaguchi prefectures, where merchants and fishermen developed paternalistic protector and protégé," or more blatantly "boss and henchman," relationships. Before the advent of modern financing and distribution systems, the merchants paid in advance for equipment, food, and other miscellaneous necessities for the fishing operations, and the fishermen paid back the debt with their hauls. Since most fishermen were indebted to the merchants, they became subordinate to them and often succumbed to exploitive treatment. The unequal relationship between the two often allowed the merchants to avoid strict weighing of the haul and drove the prices down to unreasonably low levels, while fishermen could do nothing but accept their patrons' word.[57] Japanese entrepreneurs in Hawai'i adopted this paternalistic financing system, but they weakened its feudalistic elements by replacing the arbitrary pricing with a public auction system. Unlike merchants in Japan, who often did not immediately pay fishermen for their hauls, the new fishing companies made daily payments and assumed responsibility for collecting from the dealers, who were required to make weekly payments.[58]

During the 1910s, the Japanese population expanded to more than 100,000.[59] In order to meet the growing demand for fish, the fishing companies sought to increase the catch by sponsoring competent fishermen to build their own boats and took the benefits from their active operation. This system functioned well to enable promising but penniless fishermen to raise their status and become boat owners. Heikichi Komine, who came to Hawai'i illegally from Wakayama, as previously noted, started fishing as a hired crew member on a small sampan. Then, when he obtained a good catch, a company financed a big fishing boat for him. He did not disappoint his patron and maintained a good record of hauls for the next three decades.[60]

The supportive system used by the fishing companies operated relatively smoothly and established long-lasting relationships with the fishermen because they typically earned higher wages than those in most other occupations. Moreover, the companies made substantial efforts to win the trust and loyalty of the fishermen in order to build amicable relations with them. Matsutarō Yamashiro, of the Pacific Fishing Co., for instance, helped create a *kumiai* cooperative, a mutual support organization of fishermen, and donated a $1,000 revolving fund. He also offered prize money for the boat captain with the largest catch each month and hosted a large New Year's banquet for fishermen and retailers. In the event of disaster, he dispatched a rescue boat at the company's expense.[61] Managers of other fishing companies made similar efforts to secure the loyalties of fishermen, treating them to New Year's parties, providing counseling, and, if necessary, conducting rescue operations for the victims of shipwrecks and helping bereaved families. Through their benevolence and support, the Japanese fishing companies attempted to create mutual prosperity with the fishermen.

The three fishing companies in Honolulu were located at the corner of Kekaulike and King streets, a central part of downtown Honolulu in those days. When a group of Chinese and Japanese merchants who had stalls at the Oʻahu Market opened the Aʻala Market at Queen Street and College Walk in around 1918, the Hawaii Fishing Co. and the Honolulu Fishing Co. moved there.[62] The new market included the auction houses, storage, handling, weighing, and bookkeeping services. Soon after a sampan fishing boat arrived at Kewalo Basin, usually early in the morning, the catch was transported to an auction house and displayed on the floor for inspection by wholesalers, retailers, and peddlers who regularly appeared at the morning auction. Large fish were placed side by side on the floor and put up for bidding, and small ones were grouped in small lots, prior to the beginning of the auction at around 6:00 a.m.

Even if some dealers did not buy fresh fish at the auction, they still came to check the prices and exchange information with fellow dealers.[63] Thus, the auction house functioned as a fulcrum of the fishing industry to distribute fish, exchange critical information regarding fish prices and other miscellaneous matters, and develop social networks among people in the business. In addition, the fish market and its vicinity served as a political center as well as the nucleus of social activities. Sanford B. Dole, a former president of the Republic of Hawaiʻi, described a gala at the market about 1920:

> I have called the old fish market an important institution, for not only was it the one regular market for all Honolulu and for all manner of produce, but it was Honolulu's political center where impromptu mass meetings were held and political orators held forth in election campaigns, usually each one on his rostrum of an over-turned empty salmon barrel; it was, in a way, a social center also, especially on Saturdays for then business was at its height, partly because Hawaiians made their purchases then for Sunday, the market being closed on that day, and partly because Saturday afternoon was a general half holiday and men and women dressed for the occasion. . . . Besides the traffic, there was exchange of news, some gossip, much badinage and general merriment.[64]

This observation reveals that the fish market had become an integral part of community life for business, political, and social activities in Honolulu. At the locus of the gatherings were the fishing companies that put together and distributed the harvest of Hawaiian waters.

The auction system, however, often aroused criticism and suspicion from public authorities for alleged transgressions, such as Japanese fishermen's claims that three- to-five-day-old fish were fresh, which violated sanitation regulations; dealers' collusion in bidding; and unfair price-fixing agreements.[65] The suspicions culminated in the suspension of the auction and introduced price controls by the territorial government soon after World War I. Government officials claimed that the manipulations of auction dealers were the main cause for a wartime rise in fish prices. They then set up the Kakaʻako Fishermen's Association and let it handle fish directly. This attempt immediately ran into strong opposition from fishermen, who claimed that the imposed prices were too low.[66] The criticism soon subsided because of the increase in supply and anticipated price drops, which brought into question the legitimacy of public intervention. Finally, after a brief suspension, the auction reopened.[67] Thereafter, officials did not attempt to interfere with auction houses in Hawaiʻi.

Dealer rebellion and factionalism among fishermen

The patronizing function of the Japanese companies, which won the hearts of fishermen, was not seamless enough to guarantee the satisfaction of all involved in the industry. In February 1910, twelve Japanese and eighty Chinese dealers started a boycott of fish auctioned at the Hawaii Fishing Co. Their protest was triggered by a new company policy that allowed anyone to participate in the auction. Previously, only a limited number of fish dealers and wholesalers could participate, and the sudden entry of amateurs caused a severe fluctuation in market prices.

The sense of autonomy among fish dealers led them to take bold action against the fishing company. Unlike fishermen, who were obligated to the company for its support, the dealers had far less commitment. They were customers, rather than protégés, of the company, and purchased its products to sell at leased stalls in a marketplace. In addition, the significant presence of Chinese dealers created a certain distance from the executives of the company, most of whom were Japanese. Since the Chinese had formed the Fish Dealers' Guild in 1903, their business had grown enough to allow them to build their own clubhouse on Kukui Street in 1920, calling attention to their presence in the local industry.[68] The Chinese and Japanese fish dealers agreed that amateur bidders were willing to pay more than professionals and were unnecessarily pushing up the general price of fish. Therefore, they demanded that the company either compensate them for their losses or offer an alternative sales system that would exclude amateurs.

The entry of Achu, a Chinese merchant, into the bidding also irritated the protesting dealers. Achu bought up large amounts of fish to raise the bidding price, and kept it in storage on ice to provide fish to passenger liners. His manipulation of the prices was, according to the fish dealers, the result of his conspiracy with the company.[69] Both sides refused to compromise, and the boycott dragged on. The company defended its policy, maintaining it was rational to sell to the highest bidders. The company had to pay an annual auction license fee of $600 to the territorial government, so that stopping the auction and starting an alternative sales system was its last choice. Daunted by the firm demand of the dealers, the company started direct wholesaling to secure an outlet for its products.[70]

The joint boycott of Honolulu's oldest Japanese fishing company by the Japanese and Chinese fish dealers triggered anger among the Japanese community, which harbored anti-Chinese feelings. Thus, the Japanese protesters justified their strategy by saying, "We don't need to distinguish nationality as long as we share the common interests in the same trade."[71] The strength of the transethnic tie revealed in this statement contrast strikingly with the ethnic antagonism of the plantation society, where various ethnic groups were pitted against each other. Ethnic friction among plantation workers was perpetuated by the haole (white) plantation owners, who promoted ethnic distinctions of religion and custom, and introduced a pay scale that differentiated by race and ethnicity to arouse jealousies. Through promoting ethnic antagonism and using a divide-and-conquer policy, the plantation owners discouraged the development of a unified labor force that might rebel against white control.[72] In contrast, the fisheries industry was free from such ethnic manipulations. The joint boycott as well as the aforementioned establishment and success of the Pacific Fishing Co. under Sino-Japanese collaborative management indicates that the fisheries business in Honolulu had developed a distinctive social arrangement whereby association in the same trade often outweighed ethnic antagonism as early as 1910.

The dealers' strike ended in a victory for the company. The open-access policy of the auction did not change, and the dealers had to cope with the changing situation. However, this incident conveys another important aspect of the fishing industry: the independence of the fishermen from Wakayama. During the strike, the dealers were able to continue their businesses, thanks to fish coming from the Wakayama fleet, which operated independently from the Hawaii Fishing Co. and used its own marketing channels.[73] Moreover, Wakayama fishermen distinguished themselves from others by having their own space and lifestyles. A fisherman from Okikamuro explained the difference between the two as follows:

> The members of the Wakayama skipjack fishing fleet anchor their boats at the entrance of the Nuʻuanu River, a district distant from the Kakaʻako area. . . . Generally, fishermen behave rudely and rough, but the Kishū group is terrible and they get into fights very often. But they change the valor of fighting into the energy of an outstanding job once they are out at sea. They live in leased row houses at around the Kekaulike, Hotel, and King streets. In a far from civilized manner, they sit naked in the windows of the second floor and look down the street.[74]

The Wakayama fleet anchored its sampans near Pier 16 and formed a fishing camp at nearby Honolulu Harbor and the market, whereas other Japanese fishermen used Kewalo Basin and lived in the Kakaʻako area.[75] It was around 1933 that the Wakayama fleet moved to Kewalo Basin, leaving Honolulu Harbor for cargo ships and freight vessels. The primary reason for their physical distance from other Japanese fishermen was presumably their unique fishing methods. As long as the Wakayama fleet dominated skipjack tuna fishing, it did not need to collaborate with others. Moreover, many of the Wakayama fishermen aimed at moving on to the West Coast, using Hawaiʻi as a stepping-stone for their next journey and diminishing the necessity to deepen their association with other Japanese in the trade.

This somewhat divisive distance between the Wakayama and other Japanese fishermen resulted in the creation of the Honolulu Fishing Co. in 1914, the majority of whose sampans were owned by fishermen from Wakayama. In fact, each Japanese fishing company in Honolulu had developed special ties with particular prefectures. The Hawaiʻi Fishing Co. was strongly related to those from Yamaguchi. This connection strengthened when Shinkichi Ueda, from Kaminoseki, in Yamaguchi, bought the company in 1922, changing its name to the Hawaii Suisan (fisheries) Co., and subsequently putting men from Yamaguchi in all of its executive positions.[76] In the meantime, the Pacific Fishing Co., under the leadership of Matsutarō Yamashiro, from Hiroshima, had affiliated with a larger number of fishermen from Hiroshima than rival corporations. This prefectural factionalism in the affiliations of particular fishing companies gradually diminished over time. Interestingly, Gorokichi Nakasuji, a pioneer fisherman from Wakayama, worked with the Hiroshima-dominated

Japanese fishing boats at Pier 16, Honolulu Harbor, circa 1910–1920. Photo courtesy of Hawaiʻi State Archives.

Pacific Fishing Co. during the 1920s, whereas more and more sampan owners from Hiroshima, Yamaguchi, and other prefectures came to sell their haul to the Honolulu Fishing Co. By the early 1920s, the fishermen had come to choose their affiliations with companies without exclusive consideration of the prefectural background of a company's management.[77]

The wave of backlash

Political instability in Hawaiʻi around the turn of the twentieth century, including the toppling of the Hawaiian monarchy in 1893, the subsequent switch to a Republican government, and the annexation of Hawaiʻi by the United States in 1897, caused upheaval and confusion in the customs and policies of the fisheries. Since the United States asserted government ownership of its coastal waters and extended the principle to the newly acquired territory of Hawaiʻi, US rules often conflicted with special fishing rights granted by the Hawaiian monarch. As of 1902, twenty-five cases of fishery rights were in the courts, and the Supreme Court of the territory handed down decisions invalidating the old prerogatives and supporting US government policies.[78]

With the extension of the US fisheries rule to the territory of Hawai'i, the influx of Japanese fishermen completely changed the picture of Hawaiian waters. The aggressive exploitation of sea creatures by Japanese fishing methods brought warns from scientists about potentially disastrous effects on fish as a food supply as early as 1903.[79] The preservation of natural resources gave a plausible excuse to anti-Japanese forces to counter the increase in the number of Japanese fishermen. In February 1909, territorial senator William J. Coelho from Maui introduced a measure to close the waters of the islands to noncitizens and impose a fine of $100 on violators. According to the local newspaper, *Pacific Commercial Advertiser,* the territorial Senate turned down the measure, calling it "stupid actions of the California Legislature" and in "direct violation of the treaty rights of other nationalities."[80]

Coelho's measure obviously aimed at excluding the Japanese from commercial fishing by copying the anti-Japanese measures passed in California. Since 1890, Japanese fishermen had engaged in salmon fishing and abalone diving in Monterey, California. Many of them were from Wakayama, Hiroshima, and Chiba prefectures. During the 1910s, the center of Japanese fishing moved southward, and San Pedro, south of Los Angeles, developed into the heartland of the Japanese fishing industry in California. In particular, the expansion of tuna canneries in the Los Angeles area fueled the fishing industry by opening the canned tuna market to a white population that had previously not been very familiar with eating fish. Among Japanese fishermen, those from Wakayama became prominent in tuna fishing. However, the mass appearance of Japanese on the West Coast fisheries scene provoked the state legislature into trying to eliminate Japanese fishermen by levying high taxes and imposing various limitations on their operations. The states of Washington and Oregon followed suit, passing similar laws. At the local level, certain white leaders of Monterey fisheries made an effort to loosen the grip of the Japanese by inviting fishermen of other ethnicities to fish the area, Sicilians in particular.[81]

It was, ironically, the white owners of the tuna canneries who protected the Japanese fishing rights from the hostile legislature and other legal harassment. Because the Japanese had become primary fish providers, they needed to be kept in order to secure the canneries' business interests. In exchange for the protection of these white entrepreneurs, the Japanese increased their dependence in various ways, for example, living in company-owned houses and developing the mentality of employees of the white fishery enterprises.[82] In contrast to this mutual dependency, Japanese fishing activities in Hawai'i were carried out without the protection and control of the local white elites; the Big Five oligarchy exclusively controlled the land, economy, and politics of Hawai'i, but the fishing industry was well out of their jurisdiction. The Japanese fishermen filled a sort of economic niche without relying on white capital and protection, which meant they had to protect their own rights exclusively through their own efforts.

Instead of relying on white protection, the absence of rival fishermen in Hawaiian waters and the great demand for marine products worked in favor of the preservation of Japanese fishing rights. Unlike in California, where the Japanese had rivals from other ethnic groups, in particular, Italians, Austrians, and Slovenians, only Native Hawaiians were major competitors in Hawai'i, and the native presence had dwindled sharply during the first decade of the twentieth century.[83] The fear of being unable to provide an adequate supply of fresh fish, together with references to the international treaty, crippled Coelho's attempts at eliminating the Japanese from Hawaiian waters.

To the dismay of many Japanese, however, another bill prohibiting net fishing within Hilo Bay for two years was introduced by a congressman from the island of Hawai'i and was passed into law a week later. Hilo was the largest fishing base on Hawai'i, and the new rule was expected to devastate the Suisan Co. and its Japanese fishermen. The new act, ostensibly aimed at protecting fishery resources, was viewed by the Japanese as nothing more than retaliation for the failed Coelho bill. The new rule was extremely prohibitive for skipjack tuna fishermen, for they caught live bait with nets in Hilo Bay before going out to open water. Switching from netting to alternative methods was unrealistic because it would waste their large investments in expensive nets and learn completely different fishing methods.[84] The Suisan Co. estimated that the new act would eliminate more than 250 Japanese fishermen operating in Hilo, which drove the company to petition to restore their fishing rights. When the company found its efforts to be fruitless, it hired two lawyers and filed suit in court. In the Hilo Circuit Court, company executives as well as fishermen testified and successfully appealed to abolish the prohibition.

The court victory was a blessing for everyone in the fishing industry in Hilo, but it inevitably produced negative side effects. Unexpected expenditures for legal fees severely strained the company's finances. In order to pay the fees, company executives made the painful choice to raise their commission rate from 5 percent to 10 percent on the sail haul of the boats. This new decision caused a split between the employees and affiliated fishermen into pro and con factions. With the escalation of this internal conflict, the company sued the opponents, while the latter, in turn, sued Kamezō Matsuno, the company manager, accusing him of perjury. When the dispute between the two groups appeared to have reached an impasse, the president of *Hawai Shokumin Shinbun*, a Japanese-language newspaper based in Hilo, offered his offices as a neutral negotiating site, inviting both sides to the negotiating table. Their talks broke down, and finally the opposing party left the company and formed its own organization, the Hawaii Island Fishing Co.[85]

The devastating wave of anti-Japanese legislation did not stop at breaking up Suisan Co. A new bill was approved in 1913 to ban catching *nehu* and *'iao*, live bait for skipjack tuna fishing, with nets more than twelve feet in length. The local Japanese newspaper reported that this new rule frustrated the entire Japanese fishing effort and urged Japanese community leaders in Honolulu to take daring action. Sometarō Shiba of *Hawai Shinpō*, Kinzaburō Makino of *Hawai Hōchi*, a reporter from *Nippu Jiji*, Matsutarō Yamashiro of the Pacific Fishing Co., and captains of two fishing boats, the *Tenjinmaru* and the *Takasagomaru*, embarked on a sampan, the *Kasuga-maru*, at midnight, engaged in *nehu* fishing with banned nets, and deliberately got the captain, Tsurumatsu Kida, arrested. Through the connections of these protesters with the local authorities, Kida was released and the restriction on *nehu* and *'iao* fishing was rescinded. This change in policy was limited only to the island of Oʻahu, and fishermen on the other islands were still under the ban's constraint. Yamashiro and Makino started petitioning the territorial legislature to abolish the restriction. In the meantime, fishermen on Maui sued the government in vain. Kinzaburō Makino called them "stupid" and their actions "selfish," but this case also signified that ordinary fishermen dared to stand up to protect their own rights, without waiting for the instruction of the Japanese elite.[86] The problem of the limits on catching *nehu* and *'iao* smoldered for decades without resolution. The Japanese fishing community maneuvered regulations introduced one after another by complying with new rules, or slipping through the cracks, and, if necessary, filing suits in court.

Matsujirō Ōtani challenges the Big Five oligarchy

While the arbitrary fishing regulations disturbed fishermen out at sea, the absence of clear guidelines for the sale of fresh fish continuously annoyed the merchants, especially fish peddlers. When many peddlers put fish in baskets and went out to sell, they were often stopped by the police, who imposed fines on them for sanitation violations. Such actions crippled the businesses of many. Matsujirō Ōtani, a young fish dealer from Okikamuro, challenged the unreasonable status quo; in 1910, he deliberately parked his horse-drawn wagon in front of the Department of Health building, got arrested, and brought the issue before the court. He lost his first trial case, but later won in a higher court, resulting in the legalization of fish peddling, enabling peddlers to sell their wares without fear of interference or expensive fines.

Matsujirō Ōtani carried out his struggle single-handedly and hired a lawyer at a cost of $375, which was, according to him, a "strikingly large sum of money."[87] He was, however, never a wealthy man. He obtained the funds through his hard work and financial support from a *kō* pot. Born in 1890 and raised on the small island of Okikamuro, he worked as a fisherman and ship's carpenter while waiting for a chance to go abroad. Originally, he wanted to go to Korea, as many other islanders did, but his encounter with Isojirō Kitagawa, later a pillar of the Suisan Co., who happened to be at home visiting from Hilo, changed his original plan. Lured by Kitagawa's description of the lucrative fish sales in Hawai'i, he arrived there in 1908 with little money in his pocket. After working at various jobs, he finally settled in the Kaka'ako area, where many people from Okikamuro resided, and started selling fish. In 1911, he opened a fish store from a leased space in the King Fish Market with capital from a *kō*. Because his enterprise was successful, he gradually diversified his business into the export and import of foods. In 1920, he won a bid to furnish canned crab to Armstrong Army Barracks over Theo. H. Davies and Co. and American Factors, both of which were Big Five companies.

Winning a new contract with the US Army provided great momentum for Ōtani to expand his business, a move that surprised and even humiliated his rivals. In the negotiating room, the personnel from these two companies took "extremely incomprehensible and unpleasant attitudes" toward Ōtani. In lieu of swallowing the insults, he rose up against the arrogance of the white elitists. The day following the negotiations, he called up the Davies personnel, accused them of racial dis-

Matsujirō Ōtani. Photo courtesy of Ryōko Ōtani.

crimination against the Japanese, and declared he would sever all business ties with them thereafter. He took similar actions against American Factors.[88]

Ōtani's actions were no mere reckless releases of indignation against racism. Rather, his boldness was seemingly backed by his conviction that he could succeed in business without the Big Five. In sharp contrast to Ōtani's resolution to go a step beyond a submissive minority status, Davies and Co. and American Factors retaliated by instituting racial arrangements on sugar plantations that kept Japanese workers dependent on their patriarchal control. Unlike the plantation society, the fisheries industry of Hawai'i had already developed completely different racial relations in negotiations of power, which Theo. H. Davies and Co. and American Factors failed to comprehend. After parting ways with the two companies, Ōtani successfully added the US Navy to his client list. With the buildup of US military forces in Hawai'i as a strategic point in the Pacific, Ōtani's business grew. Besides running his own company, he joined the management of the newly established Hawaii Suisan Co., a successor to the Hawaii Fishing Co., with other entrepreneurs from Yamaguchi, and became one of the leaders of the fishing industry in Hawai'i.

Japanese fisheries reach their full glory

> We should say that fishing in Hawai'i is exclusively in the hands of Japanese. Although there are fishermen of other ethnicities, they are far from our major competitors.... In other islands of the archipelago, the Japanese domination of fishing remains the same as in Honolulu.[89]

This statement, made by the Japanese Chamber of Commerce in Honolulu in 1922, was far from extraordinary. The Japanese in Hawaiian waters dominated operations and successfully developed the local fishing business into the third-leading local industry, following sugar and pineapple production, by the early 1920s.

The fishing companies were the backbone of Japanese hegemony in the industry. Running fishing companies and exclusively buying fish catches in Honolulu and Hilo, both of which were core areas of fisheries in the archipelago, Japanese fish merchants established their control of the general seafood supply. Because these companies served as primary financiers to Japanese fishermen, their growth had multiple effects on the expansion of Japanese fishing fleets and the improvement of their operations. As discussed in this chapter, the process of enlarging Japanese presence in Hawaiian waters included harmonious as well as frictional encounters with native fishermen and rivalries as well as cooperation with Chinese dealers. Although the standing of natives was noticeably diminished in the seas of Hawai'i, they still remained part of the industry by operating independently or working together as hired crews on Japanese sampan boats. Some of the Chinese had become integrated into the management of Japanese-dominated fishing companies, keeping a fair bit of influence in the wholesaling and retailing of fresh fish in the islands.

In the meantime, the dialogue between the Japanese and the haole elite was embroidered with far more complicated and inconsistent colors. That the Japanese had fearlessly challenged the white hegemony in local politics and the economy reflected their self-determination, and it also meant the absence of the white protection and patronage enjoyed by their counterparts on the West Coast. The potential for friction with white authorities in Hawai'i continually lurked beneath the surface of the islands' ocean industries and came to light again and again, as indicated by the introduction of fishing regulations aimed at curbing Japanese operations.

Nevertheless, these prohibitive elements could not stop the rise of the Japanese in the seascapes of Hawaiʻi after the turn of the twentieth century. Playing a vital role in providing the harvest of the sea to local consumers, the Japanese had successfully built a solid foundation for Hawaiʻi's fisheries by the late 1920s.

3 The Heyday of the Japanese Fishing Industry in Hawai'i

The expansion of fisheries and the birth of Hawaiian Tuna Packers

The growth of Japanese fishing operations coincided with the expansion of fishing-related industries, such as the sales of fishing gear, bait, fuel, water, and food for crews. In those days, Japanese fishing gear merchants imported hooks, longlines, bamboo poles, and other maritime hardware from their hometowns in Japan, as well as from the continental United States, and modified them to fit the fishing style of Hawai'i.[1] Ice was indispensable for the preservation of perishable aquatic products in Hawai'i's warm temperatures, and the Japanese had started an ice-making business in order to supply ice to fishing boats, auction houses, and fish market stalls.

The boat-building business flourished in Hawai'i. Fishing and boat building were specialized, separate occupations, and fishermen usually were not boat builders. It was Japanese boat carpenters who came to Hawai'i and engaged in ship building. With a few exceptions, such as Gorokichi Nakasuji, who brought his own boat from Japan, Japanese fishermen usually used boats built in Hawai'i. The design of these boats, or sampans, was basically Japanese and was modified as time went on. Generally about twenty feet in length soon after the turn of the twentieth century, sampans were used principally for line and gillnet fishing.[2] During the 1910s, sampans were equipped with gasoline engines, rapidly replacing smaller ones powered by sails or sculls and enabling fishermen to venture much farther out to sea than the Native Hawaiians, who seldom went beyond the coral reefs. With the use of engines, boat lengths increased from thirty to sixty feet. During the 1920s, Kiyoshi Kashiwabara installed a diesel-powered engine. Since diesel fuel was more than 50 percent cheaper, others followed his example. After the introduction of the diesel engine, skipjack tuna-fishing boats increased in size, some of them ninety feet in length, with 160-horsepower engines.

The boats used for catching bottom fish were smaller, but improved refrigeration systems with ice enabled these boats to preserve fish longer and stay out for weeks at a time. Japanese carpenters who maintained yards for the construction and repair of fishing boats in Hilo and the Kakaʻako district of Honolulu constantly promoted innovations and technological advances.[3] They transformed sampans, which were originally Yamato-gata-style Japanese fishing vessels, into those designed for wider-ranging offshore operations in rough, open waters by adding high bows and stable hulls.[4]

The increasing supply of fish sparked the rise of seafood processing. Matusjirō Ōtani started *kamaboko,* a kind of fish pâté or steamed *surimi,* production, and invited a technician from Hiroshima to work with him. Thanks to his leadership, Ōtani's products received a favorable reception and his office had a rush of orders from all parts of the archipelago.[5] In the meantime, Aratarō Yamamoto, born in Tanami, in Wakayama, and educated at the National Fisheries Institute in Tokyo, came to Hawaiʻi in 1906, and started *katsuobushi* production. In Japan, skipjack tuna was eaten raw, boiled, or roasted, but it was most valuable when processed into *katsuobushi,* an essential ingredient of Japanese cooking. Skipjack tuna is soft fleshed and decomposes rapidly in hot weather, which coincides with the period of peak catches, and the Japanese had developed its production since the early stages of skipjack tuna fishing. Yamamoto brought the manufacturing process, including boiling, smoke drying, and molding, into Hawaiʻi. But the natural conditions were different from those in Japan, and harmful insects in the archipelago damaged the product. After a continuous process of trial and error, he finally succeeded in producing high-quality *katsuobushi* and expanded the market for the islands. Together with sugar and pineapples, *katsuobushi* made in Hawaiʻi had become one of the most favored *omiyage* gifts to take to Japan.[6]

The birth of a tuna cannery during the later 1910s furthered the distribution of fish, especially canned skipjack tuna. F. Walter Macfarlane, a young pineapple planter, established a cannery at Cook Street in Kakaʻako. Later, he added a shipyard, hired Japanese craftsmen, and started building fishing boats equipped with gasoline engines. Because he lacked a working knowledge of the seafood industry, he invited professionals in canned tuna production to come to Hawai'i from the mainland, with the lure of high salaries. He also introduced an innovative system of paying fishermen monthly salaries, but it merely increased personnel expenses and worsened the financial condition of the company. Finally, Macfarlane sold the cannery to American Factors, but operations under the new management did not go well. Only slightly more than a year after the takeover, a group of American and Japanese entrepreneurs and fishermen purchased all the facilities at a low price and restarted the company as Hawaiian Tuna Packers Ltd. in 1922. E. C. Winston became the first president of the company; Matsutarō Yamashiro of the Pacific Fishing Co. assumed the vice presidency; his eldest son, Matsuichi, took the post of secretary; and Tsurumatsu Kida became a director. Under the joint management of Americans and Japanese with rich experience in fisheries, the reorganized tuna cannery markedly increased its productivity and sales by selling to outside markets; in 1927, it produced about 20,000 cans, and the number expanded to 150,000 in 1930.[7] Just after the birth of Hawaiian Tuna Packers Ltd., Suisan Co. started cannery production in Hilo.[8]

The soaring demand for fresh fish for canneries required the hiring of more skilled fishermen. Yamashiro and Kida went to Wakayama and recruited thirty-seven experienced fishermen and their families to move to Hawai'i.[9] Other fishermen from Wakayama also invited their colleagues from home, and the size of the skipjack tuna fleet in Hawai'i rapidly increased.[10] According to a survey conducted by the Japanese Consulate in 1924, 90 percent of the 1,124 immigrants from Wakayama engaged in fishing, accounting for 38 percent of O'ahu's fishermen.[11] To facilitate production, Hawaiian Tuna Packers introduced a system among Japanese fishing companies that financed skilled fishermen to build their own sampans in exchange for exclusive rights to the catch. It deepened its relationship with the Japanese fishing companies by agreeing to market fish and other sea creatures caught accidentally by cannery-affiliated sampans. The fishing companies often reciprocated by providing skipjack tuna to the cannery whenever the cannery was short of fish. Through this mutual exchange of fish and support, the two organizations worked collaboratively as the driving force in the development of commercial fishing in Hawai'i.

Fishing styles and methods

At the turn of the twentieth century, an insignificant number of Japanese fishermen operated in Hawai'i. By the late 1920s, the total number of Japanese fishermen had increased to slightly less than 1,100. Fishing companies in Honolulu and Hilo owned 90 percent of all registered engine-powered fishing boats in Hawai'i and landed fish worth $1.2 million annually. If fish provided by individual boats and fishponds were included, the total price would have been close to $2 million. These reported earnings from fishing were usually three to four times the total of sugar plantation wages during the 1920s and 1930s. Although the annual income was commensurate with experience and significantly varied according to the level of each fisherman's skill and the acumen of boat captains, an average catch in those days reached $2,000 to $3,000 for a small sampan accommodating one to two people, $10,000 for a midsize catch with a crew of three to five, and $20,000 for a large boat with seven to ten fishermen.[12] The fishermen usually worked on a shared basis. For example, if a boat had a haul worth $1,000 at auction price, a fishing company took 10 percent ($100), and an additional $300 for various expenses, such as water, food, fuel, and so on, from the original $1,000. If the boat owner took 50 percent commission from the remaining total of $600, the crew members would share the $300 balance evenly ($100 per member). If the owner was also a part of the crew, he took both the commission and part of the remaining share.[13]

Japanese fishermen imported a harsh style of training novices from their home country. When Walter Asari, the son of a fisherman from Wakayama, was old enough to go to high school, he started working on his father's skipjack tuna boat. As the lowest-ranking member of the crew, he cooked meals with a *hibachi* brazier, washed the crew's uniforms, and unloaded fish from the hold. Asari said:

> Even if we had a big catch and were unloading thirty- to forty-thousand pounds (of skipjack tuna), and the hold was full of fish, one person had to throw all of it up on deck. When I was doing it, nobody would think of helping me. They were really strict. In fact, the old-timers would stand on the dock and tell me to hurry it up.[14]

Katsukichi Kida received similar training from his father, Tsurumatsu. When he started working as a school-age boy, Katsukichi was assigned to many dirty jobs, and corporal punishment was not unusual in his everyday life on the boat.[15] In the old days, "the old fishermen wouldn't tell you anything. If you made a mistake, the bamboo poles would come out flying," he said.[16] By doing hard, dirty jobs, a young novice was expected to observe and learn fishing skills from the veterans.

Once he moved up the shipboard hierarchy, from being a cook to a full-fledged fisherman, he aimed at stepping up to the next stage, which was to own his own boat and become its captain. Unlike fishing in the old days, when people went out to sea with small sailboats, the installation of engines and increasing size boosted the price of sampans. A gasoline-powered, bottom-fish sampan with 65 to 75 horsepower cost about $10,000, and a skipjack fishing boat was priced at about $20,000 to $30,000.[17] With the introduction of the diesel-powered engine, boat construction costs became even higher. Because it was a large investment to build a vessel, many fishermen sought the patronage of affiliated fishing companies. Occasionally, a boat builder gave advances to his customers. Seiichi Funai, a shipyard owner from Susami in Wakayama, built boats for those who longed for but could not afford to buy their own. Later, his customers repaid him from their catch.[18]

Seiichi Funai came to Honolulu in 1917 in a sailboat carrying a cargo of *shoyu* (soy sauce). When the boat captain went back to Japan, leaving Funai and two other crew members behind, they obtained permission to stay in Hawai'i through the good graces of the Japanese consul general.[19] Later, Funai told his eldest son, Teruo, that a shipwreck forced him to stay, but Teruo believes his father ended up in Hawai'i because he met his mother, Kimi, a nisei (second-generation) Japanese woman born on Maui.[20] Whatever the reason, Seiichi Funai settled in Hawai'i with help from others from Susami, who assisted him financially in starting up a shipyard in Kaka'ako.[21] Before his retirement in the mid-1950s, he built more than 150 sampans.[22] The boats were products of his excellent craftsmanship and mutual support from fellow fishermen, fishing companies, and boat builders in his area. From 1924 to 1931, the number of sampans in the Hawaiian archipelago expanded from 272 to 355, marking an increase of 30.5 percent in just seven years.[23] Their sizes varied from single-person, 3- to 4-horsepower boats to multiperson, 200-horsepower boats. The owners of the boats were almost exclusively Japanese.

Japanese fishing operations in Hawai'i during the 1920s could be categorized as several types: pole-and-line for skipjack tuna (*aku*), longline for yellowfin tuna ('*ahi*), and net fishing. The *aku* and '*ahi* boats were similar, although their designs changed and improved over the years.[24]

Seiichi Funai. Born in Wakayama, Funai came to Hawai'i in 1917 and built more than 150 sampan fishing boats, including the *Kinan-maru*. Photo courtesy of Teruo Funai.

Usually, the largest boats were engaged in skipjack tuna fishing, since they were equipped with many hatches into which seawater constantly flowed. In the late 1920s, about ten to eleven crew members manned each boat. At first, they netted *nehu* (anchovy) and *'iao* (silverside) in the shallow waters of the harbors, streams, and protected reef areas and transferred them to the boat hatches. Some boats engaged in day baiting, which usually started at dawn and ended when sufficient bait was captured, but most bait fishing was done at night. Because bait gathered by lamplight attracted mosquitoes, fishermen had to protect themselves from mosquito bites by wearing rice bags. Many of them could not endure the fierce mosquito attacks, giving up their operations in Hawai'i and moving to the mainland.[25]

After the nighttime ordeal of bait fishing, they headed to the fishing spots in the morning to search for schools of skipjack tuna. In the beginning, skipjack tuna fishing was done with sailboats, and fishermen caught only inshore fish weighing about four to five pounds. With the introduction of the engine after about 1911, they started going farther and farther from the coast for better fishing sites and bigger catches weighing as much as twenty-five pounds per fish.[26] Fishermen searched for fish by scanning the horizon for bird flocks because schools of skipjack tuna were usually accompanied by seabirds flocking and feeding on prey driven to the surface by the fish. Wearing denim kimonos and straw hats to protect themselves from the sun and flying hooks, each crew member equipped himself with a long bamboo pole about twelve feet in length and two inches in diameter with a strong line about ten feet long attached to its tip. Once a school was sighted, they took out live bait from the hatches and threw it into the sea to attract the school of skipjack tuna to the surface. Ocean spray was kicked up with a *shakushi* (Japanese rice paddle) in this method of fishing, to camouflage the fishermen standing along the stern and sides of the boat. They then began to dip their lines into the sea, pulling up the fish with barbless steel hooks, holding each fish in their arms and removing the hooks. They continued this cycle until the fish supply was exhausted. Mastering this art of fishing took considerable practice and experience.

In Japan, fishermen usually fished from the side decks of boats, but such an arrangement would allow the big waves of Hawai'i to easily swallow up the boat and its crew. Therefore, fishermen and boat carpenters in Hawai'i adopted a new design in which fishermen were able to catch fish from the stern and were less vulnerable to the waves. Once they finished fishing, they went back to port as soon as possible. Because they did not ice the catch, they kept the fish in bait wells.[27] Skipjack tuna were caught throughout the year, but were remarkably migratory and seasonally abundant, more so in the summer than in the winter.[28] Pole-and-line skipjack fishing was done the same way for a long time, with only the minor technological advance of partial conversion to fiberglass poles after World War II.[29]

Fishermen from Yamaguchi, Hiroshima, and other prefectures employed methods of longline and net fishing. Usually, the owner of a small fishing boat engaged in shallow-water fishing using nets, hooks, and lines. But in Hawai'i, the reefs hindered the use of draw nets. Instead, cast nets were usually used. In order to catch the *opelu* (mackerel, *Decapterus pin-*

Skipjack tuna fishing before World War II. Fishermen wore straw hats and denim kimonos, and used bamboo poles to fish for skipjack tuna. Photo courtesy of Hawai'i State Archives.

nulatus), for instance, they threw each net into the sea and let it sink to about seven fathoms. Then, they dropped the bait, which was mashed pumpkin wrapped in a cloth, and suddenly raised it up in jerks to unwrap and scatter it to attract the fish. While the mackerel were eating the bait, they pulled up the net. The wrapped bait was also used for catching expensive deep-sea fish, such as *ulaula* (red snapper, *Etelis carbunculus*),

Skipjack tuna fishing before World War II. Photo courtesy of Hawai'i State Archives.

mahimahi (dolphinfish, *Coryphaena hippurus*), and *ulua* (crevalle, *Caranx*). As the Hawaiians used the pumpkin bait for *opelu* fishing before the arrival of the Japanese, the latter presumably adopted this method from the former and refined the nets.[30] The longline methods were also used for bottom fishing. Fishermen sank the main line to the bottom using stones together with branches. In the late 1920s, the line was about 1,000 to 1,500 fathoms long and each branch was about 4.5 fathoms long. The bait was live *opelu* or *akule* (bigeye scad, *Trachurops crumenophthalmus*).[31]

The longline method of catching *'ahi* tuna (*shibi nawa*), with a relatively large vessel equipped with ice containers, was Hawai'i's second-largest fishery.[32] After it was introduced by Nakasuji Gorokichi, Japanese from Wakayama, Yamaguchi, Hiroshima, and other prefectures engaged in it. They used longlines with buoys and flags. The main line was, according to observations of the late 1920s, the size of a pencil in diameter and about seven miles long, with around 150 branch lines. The boat headed to the fishing grounds early in the morning. The fishermen located a favorable area by carefully observing the speed and direction of the current, and started to lay the lines in a semicircle. The main line was kept afloat on the surface by buoys, to which a thin bamboo pole with a red or white flag was attached at the tip. Below each flag, a longline was dropped to the depth where the fish were supposed to be (300 to 500 feet).

Because the red and white flags were placed alternately, the fishermen could see when and at what point a fish bit and pulled down the flag to which the branch line was attached; they immediately went to the spot and pulled up the fish, a process that took about two hours. After this, they anchored near the shore. The next morning, they went back to the same area and repeated the operation. The tuna boats were comparatively small compared to the skipjack tuna boats, since the size of a skipjack tuna catch usually exceeded that of an ‘ahi tuna catch. Unlike small boats, which went out early in the morning and came back at sunset, larger ones were equipped with ice containers and often made longer trips, remaining out at sea for days. After 1920, tuna-fishing boats from Honolulu usually ventured close to Wai‘anae and Kahuku on O‘ahu, but larger, deep-sea sampans traveled hundreds of miles from Hawai‘i and stayed two weeks or longer, until they had a catch large enough to justify returning to port.[33]

Community life and women's contributions to the fisheries

"Most of the fishermen were nice. There were some arrogant men. But such fishermen had large families, so they worked hard," said Akira Ōtani.[34] Born in 1921, the second son of Matsujirō Ōtani, Akira grew up in Kaka‘ako. In his childhood memories, residents of the community, both men and women, old and young, always helped each other and lived in a close network of families and locales. When a sampan came back to Kewalo Basin, which was right next to the Kaka‘ako district, children swarmed around it, shouting "*okazu* (dish)," and were given a free share of the catch. "Fishermen were generous," as Teruo Funai later recalled. Teruo was born in 1926. He also remembers that his father, Seiichi, a boat carpenter, never received free fish. Through everyday contact with fishermen, he sensed how tough their jobs were and hesitated to take advantage of their goodwill.[35]

Akira Ōtani and Teruo Funai's stories of mutual support and concern might be repeated in other Japanese fishing communities in the Hawaiian archipelago. Most people whose lives were closely associated with the ocean clustered at fishing ports, which developed into several major fishing towns

Kewalo Basin in 1935. Photo courtesy of Hawai'i State Archives.

in the islands. On the island of Oʻahu, Kakaʻako was the largest fishing community. Outside of Honolulu, Waiʻanae attracted more than a dozen fishermen from Hiroshima. On the island of Hawaiʻi, the Japanese fishing population was concentrated in Hilo, while the island of Kauaʻi had several small communities, such as Kukuiʻula, Wahiawa, Hanapēpē, Waimea, Pākalā, and ʻEleʻele. The south coast of Maui, facing the islands of Molokaʻi, Lānaʻi, and Kahoʻolawe attracted Japanese fishermen who gathered around Lahaina, Māʻalaea, and Kīhei.[36] Many fishermen in Honolulu and Hilo operated with larger sampans, while those in other towns owned small boats and sold their catch to the local population.

Matsutarō Shimizu came to Hawaiʻi in 1921, from Tanabe, in Wakayama Prefecture, at the request of Isematsu Takenaka, a skipjack tuna fisherman who originated from Haya (now Tanabe city) in Wakayama. Takenaka also invited his relatives and fishermen friends to Hawaiʻi and added them to the Wakayama fleet. His wife, Haru, came to Honolulu from Haya to join her husband in 1920 and gave birth to a daughter, Shizue, and two sons, Tokiharu and Yoshiharu. However, the Takenaka

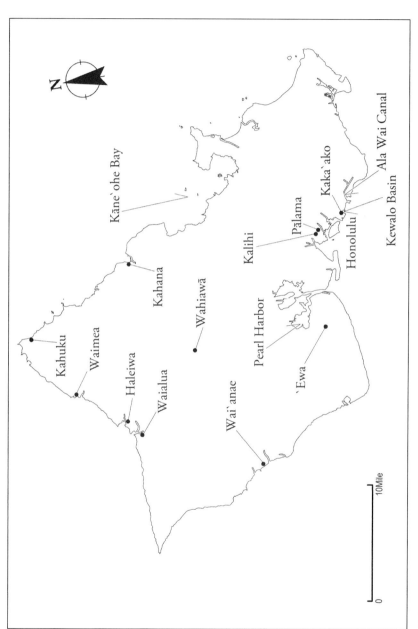

O'ahu.

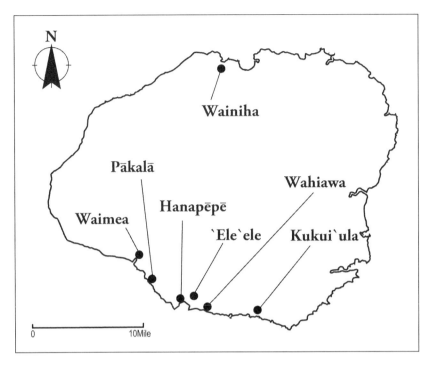

Kaua'i.

family in Kakaʻako did not last long because Isematsu died, leaving his wife and three children. Later, Haru remarried, to Matsutarō Shimizu, who was raising a son as a single father. Shizue remembers that her stepfather, Matsutarō, always stayed out at sea because "he was a fisherman by nature," she says.

> He did not care about coming home until filling his boat with his catch to full capacity, and he continued his work in open water even during storms. He never knew what giving up meant, always being *shinbō, shinbō* [patient, patient]. Only when fuel ran out, he returned to the port saying, "*shōganai*" [it cannot be helped].[37]

Even when Matsutarō landed, he rarely stayed home. Instead, he went to the port and spent most of his time taking care of his fishing gear and sampan. The chronic absence of fathers was quite common in fishermen's households.

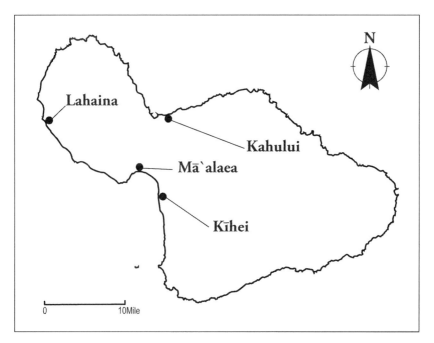

Maui.

> Children of fishermen rarely saw their fathers because they did not come home very often. Even when they were at home, it was only for a couple of days. They usually left home very early in the morning before the children woke up, so the children knew their mothers' faces, but didn't know their fathers'.[38]

Such a pattern of family life spawned social and economic arrangements specific to the fishing community. While fishermen stayed out at sea, only a few wives chose to stay home and became full-time housewives; many of them found themselves working in manufacturing industries. In Honolulu, Hawaiian Tuna Packers absorbed considerable numbers of fishermen's wives and daughters. They mainly worked splitting the fish lengthwise and sorting the meat.[39] The extensive women's employment at the manufacturing industry was customary at shore localities back in Japan where women primarily took part in processing of the catch when their husbands and fathers went to fish. The large presence of Japanese women at canneries in Hawai'i was, therefore, a customary gender role they brought from Japan.

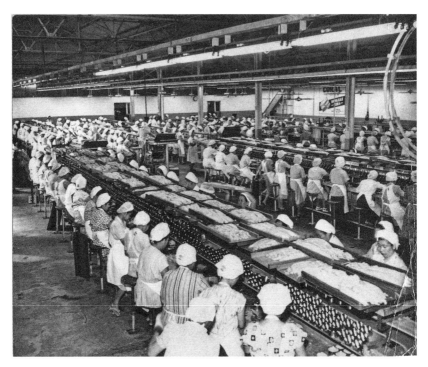

Assembly lines at Hawaiian Tuna Packers in 1959. Many wives and daughters of Japanese fishermen worked at the cannery. Photo courtesy of Hawai'i State Archives.

The involvement of Japanese women in manufacturing was eminent in the Pacific Coast canned salmon industry. In Steveston, Canada, for instance, the men operated the cannery boats, and the women worked mostly as slimers and fillers at a salmon cannery. Behind the mobilization of Japanese women was the agreement of contractors and owners that wives and daughters from the families of fishermen could be paid less than men.[40] Japanese women also took for granted that they would go to work as soon as they landed in Canada. Moto Suzuki, a fisherman's wife, stated, "When I came [1925] I was almost six months pregnant. It was in July, fishing season, so I started work in the cannery almost right away."[41]

In Steveston, the presence of European Americans, Native Americans, and Chinese contractors often spawned a competitive atmosphere and created ethnic divisions and tensions. However, such problems were much smaller in Hawaiian canneries.[42] Hawai'i had different patterns of racial and ethnic distribution, and about 87 percent of the female workers in

Hawaiian Tuna Packers. Photo courtesy of Hawai'i State Archives.

manufacturing, including tuna canneries, were Japanese. The homogeneity reduced ethnic tensions at plants to an insignificant level.[43] When Tsuru Yamauchi, an Okinawan woman, started working at the tuna factory because "they needed as many people as possible during fish season," she found that women, mostly Japanese, were cutting, cleaning, and canning fish. All the forewomen were Japanese and the language spoken at the cannery was Japanese.[44] The women received 20 cents per hour for skinning the fish, a modest wage. Moreover, the offensive fish odors at the cannery annoyed Yamauchi, a housekeeper without previous working experience at a fishery-related plant. She said, "I hated the smell. I brought clothes to change into, but I didn't have too many clothes. It could not be helped. It smelled so bad that I could not walk in front of people."[45]

Nevertheless, Shizue Shimizu remembers that "all of the fishermen's wives in Kakaʻako went to the Tuna Packers for work. Their workload was not very hard. The work shift was from 7:30 a.m. to 4:00 p.m., and there was a thirty-minute lunch break."[46] Her recollection conveys that the canners hired Japanese women at low wages to work long hours in a smelly environment during the high season. During the low season of skipjack tuna fishing, there were weeks when only a few hours of work were available. This seasonality lowered the average pay rate. In 1939, a woman's weekly earnings in miscellaneous manufacturing groups, including the tuna canneries, was $4.40, much lower than that in the pineapple canneries, which was $13.40 during the same period.[47]

Fluctuating work hours and low wages did not deter women from fishering families from engaging in manufacturing work. Besides the fact that even meager earnings were indispensable for swelling the family budget, women flocked to the canneries and other workplaces to find relief from the loneliness caused by the absence of their husbands and to share other difficulties of their daily lives. The massive involvement of Japanese women in a gender-divided workforce soon produced a subculture in which work functioned as their social center, being together and chatting with coworkers. Because most of these women did not have extended family members on whom they could depend, coworkers always helped each other with various aspects of everyday life.

Finding a balance between working outside and doing household chores, including caring for children, had always been a major challenge for working women. Unlike the cane fields, where there were day care facilities for working mothers, the Kakaʻako fishing community did not have provisions for child care. Women then developed social networks, which easily cut across household, occupation, and neighborhood, and reduced the burden of mothering. When Teruo Funai was a baby, his neighbor took care of him when his mother, Kimi, was busy managing a large family, including her four siblings and at least six apprentice workers of her husband's shipyard.[48] When Akira Ōtani's father Matsujirō—an "extraordinarily stern and business-oriented man" in the eyes of Akira's neighbor and friend Teruo Funai—started peddling fish, his mother, Kane, constantly supported him by working at pineapple and tuna canneries while raising eight children. When Matsujirō came back late at night from peddling with his horse-drawn wagon, Kane greeted him, fed his horse, and washed his muddy carriage. When one of their sons was seriously ill in bed, Kane encouraged her hesitating husband to go on a trip to San Francisco and Seattle to make new business deals. "I sincerely appreciate her for her great contribution," Matsujirō later said of Kane's long-term support and devotion to the family and his work.[49] Kane's extraordinary hard work is, of course, worthy of great praise, but support from her mother, who lived next door, and a Japanese woman she hired as a babysitter should

Japanese women chatting on River Street in downtown Honolulu, 1930. Photo courtesy of Hawai'i State Archives.

not be omitted from the Ōtani's family history. In addition, the watchful eyes of a closely knit neighborhood, coupled with very light automobile traffic, further reduced the anxiety of child care. Akira Ōtani always enjoyed playing baseball on the street with his friends, without the fear of being hit by a car or dealing with crime. He ran away only when he hit a home run that he thought damaged someone's home.[50]

On top of their work in canneries and stabilizing community life through developing friendships and neighborhood ties, Japanese women made significant contributions to the fish business. Since the dawn of Japanese fishing in Hawai'i, they stood at a fulcrum between fishermen and consumers by peddling their husbands' catch at plantations. Even after the establishment of the modern market system, they remained deeply embedded in local distribution channels as peddlers actively interacting with consumers, sellers of seafood at fish market stalls, and clerical workers in traders' back offices. Lucy Robello, a Portuguese American woman born in 1905 and raised in Waialua on O'ahu, gave an interesting account of these fish peddlers:

> And anytime the fish peddlers—mostly Japanese women—would come, we had fish, because Portuguese like fish. The peddlers had the fish box in the back with chunks of ice. They would put 'em in a scale and you would buy what you wanted. Mostly [it] was *akule* [bigeye or goggle-eye scad], *'ōpelu* [mackerel scad], little fishes like that. If it was a reasonable buy, my mother bought plenty and we'd have fresh fish that day. We had no refrigeration, so the rest was salted. When the Japanese peddlers didn't have fish, they'd come with head cabbages and sweet potatoes, and we'd buy them.[51]

This account reveals that the custom of Japanese women's fish trading spread beyond their own ethnic niche to the dinner tables of other fish-eating ethnic groups. With the sales style they brought from Japan, such as preserving perishable items with ice and substituting vegetables for fish when the latter was not available, they aggressively expanded this new market, selling their wares door to door to an ethnically diverse population in Hawai'i, striving to meet a variety of consumer demands.

In downtown Honolulu, wholesaling and retailing firms had the distinct characteristics of family-run businesses in Japan, where the women's presence was critical to run the business smoothly, with their knowledge of personal networking, influencing strategic decisions, personnel policies, and financial management.[52] As in many family firms in Japan that depended on women as their corporate officers and registered principals,

Japanese women peddling. Photo courtesy of Hawai'i State Archives.

Matsujiro Ōtani made his daughters, Florence and Gladys, staff members of his company, together with his sons, Jiroichi and Akira.[53] With the encouragement of his children, Matsujirō took over Aʻala Market from Mr. Ahoi, a Chinese man, and leased the land the marketplace had occupied for twenty-five years from the Dillingham Corporation and its subsidiary, the Oahu Railway and Land Co., Ltd., in March 1939. In the following year, a disastrous fire occurred, causing $100,000 worth of damage. Undaunted by the catastrophe, Ōtani decided to undertake not only reconstruction but additions, expansions, and improvement of their facilities, totaling $160,000. Behind the emergence of a new and finer market center from the ashes were the support and contribution of his sons and daughters. Although Matsujirō was hospitalized almost every year due to overwork and tremendous pressure to secure financing from banks that were unwilling to finance him, the family weathered these difficulties by working together and enabled the renovation and construction of a new marketplace.[54] The local newspaper, the *Honolulu Star-Bulletin*, wrote that Matsujiro Ōtani, Ltd., had been conducting business generally using "family bases," with its officers largely his sons and daughters.[55]

The involvement of women in the operation of Ōtani's enterprises was far from exceptional in a family business in Hawaiʻi. A survey published by

Ōtani at work. Matsujirō Ōtani's children supported his business. From the left, Matsujirō; Gladys, his second daughter; Jiroichi, his eldest son. At the far right, Florence, his eldest daughter. Photo courtesy of Akira Ōtani.

the Women's Bureau revealed that, as of 1939, women were more than 62 percent of all employees in Honolulu's retail market, and 46 percent in Hilo and other areas.[56] Women's participation in the retail business, regardless of sales items, was thus significant in Hawai'i. Although gender-specific statistics of fish wholesalers and retailers are not available, anthropological research conducted in the early 1970s traced the decades-long, highly significant presence of women in fish dealing, which required "a little capital and a lot of work . . . work a ten to fourteen-hour day, seven days a week."[57] Many of these women were of Japanese descent. Whether they dealt directly with customers or confined themselves to seemingly supporting roles as clerical workers, women occupied key positions in the business of the marketplace.

Personal networking through religious activities

> Wearing only a *kanaka* [Native Hawaiian] shirt, a wife of a fisherman easily goes anywhere. She loudly speaks with her frank dialect without the slightest hesitation, and does not care a bit even if money runs out because of

storms lasting for days. She does not mind eating only rice gruel with tea flavor for five to ten days. Her earnest concern about the safety of her husband is, however, uncompromisingly strong. Buddhist temples and Shinto shrines receive donations from quite a few fishermen's wives. They also willingly make generous relief donations.[58]

This character portrayal of a fisherman's wife appeared in a local Japanese newspaper in 1910. Above rank and frugal disposition, it emphasizes the ardent piety of fishermen's wives. The dangers inherent in fishing and navigation compelled them to seek divine protection. The women's network of help and communication significantly reduced their sense of chronic loneliness and anxiety for the safe return of their husbands. Nevertheless, their souls still had empty spaces, which only religious practices could fill.

Patron deities were centrally placed in fishing communities in Japan, where men and women engaged in a variety of rituals to ensure the safety of crews, to pray and give thanks for good hauls, and to propitiate the deities that control the seas and the destinies of fishing communities.[59] When the Japanese came to Hawai'i, they brought their beliefs and practices, which functioned as glue in the social fabric. Hawai'i already had several religious sects of Buddhism and Christianity, both of which took deep root in local Japanese society. But Kotohira Shrine was the most revered as the guardian deity of fishing, navigation, and commerce among sea people.

In Japan, the Kotohira-gū [shrine], which was popularly known as Konpira-san, was located in Kagawa Prefecture facing the Seto Inland Sea, where local seamen had worshiped the deity as the protector of navigation for dozens of centuries. The Kotohira-gū, standing at the top of a mountain in the middle of flat land, had served as an important landmark for ships sailing off the coast. Gradually, as a natural beacon that saved numerous lives, it had won a reputation as the guardian of navigation among coastal communities and attracted people from all over Japan. The wave of modernization sweeping the nation since the Meiji Restoration in 1868 did not wash away the popularity of Kotohira-gū.

The history of Kotohira Shrine in Hawai'i began with the appearance of Japanese fishermen on Hawaiian shores. As early as 1901, Wailuku on the island of Maui had already seen its birth.[60] In Honolulu, the Fishermen's Association set up a household shinto altar in its office in Kaka'ako and offered prayers to Kotohira Shrine beginning in 1919. This association, formed by sampan fishing boat owners and run mainly with their contributions, regularly discussed various matters regarding fishing operations and prices; in the event of a disaster, the association sent out search parties at its own cost. It was a natural consequence that the Fishermen's Association chose the god Kotohira to show their reverence and adoration. In 1921, the altar at the association office developed into an independent shrine. It acted as an emotional icon for Kaka'ako fishermen and their family members until it disappeared in 1946.

Honolulu had another Kotohira Shrine at the corner of Walter Lane and North King Street in Kapalama, near the downtown area. Rev. Hitoshi Hirota from Hiba-gun (now Shōbara city), a mountainous region in central Hiroshima Prefecture, erected it together with his friends in around 1920. In 1931, it was moved to a new location at 1045 Kama Lane. This renewed shrine stood on a piece of property measuring 50,075 square feet, purchased with $25,000 in donations from individuals and community members. The shrine boasted the largest total space in all of the precincts, making it the largest of all the Shinto shrines in Hawai'i, until it lost two-thirds of its land in 1957 due to the construction of the Lunalilo Freeway.[61] The grand scale of the Kotohira Shrine embodied its wide support from fishing communities among and outside of O'ahu, and reflected the prosperity of the Japanese fishing industry during that time. Besides Kotohira shrines, O'ahu had other guardian deities ensuring safety at sea, although their popularity was not as great as that of Kotohira; Ebisu, a god of prosperity and traffic safety, was enshrined in Hale'iwa, 'Aiea, and Kaka'ako, to ensure a rich haul and safe navigation.[62]

Although they visited the Kotohira Shrine and other religious places in Hawai'i, people sought divine grace in their respective home villages in Japan and actively donated to them. Hakuseiji Temple and Ebisu Shrine in Okikamuro, for instance, commemorate many islanders who sent money from abroad, including Hawai'i. The magnificent main hall of Hakuseiji Temple and the fine stone steps of the Ebisu Shrine were among the fruits of their devotion. The *kō* organizations, nominally religious pilgrimage groups that focused devotional and recreational energies on particular shrines, temples, and other guardian deities symbolizing specific graces, also functioned as religious as well as social centers of the community. As of 1931, people from Okikamuro in Hawai'i formed nine *kō* organizations, in which both men and women participated. Some of these *kō* groups drew more male than female members, and vice versa; the Kannon (goddess of mercy) *kō* attracted more female participants than men at

Ebisu Shrine at Okikamuro. The stone steps of the shrine carry the names of donors. The biggest stone pillar reads, "Kitagawa Isojirō of Hilo." Kitagawa was a leader of the fishing community in Hilo. Photo by the author.

its monthly gatherings, and the Hachiman (god of war, protector of nation and people) *kō* gathered larger donations from men than from women, and sent money to the bereaved families of soldiers in Okikamuro. Regardless of different gender balances, these *kō* groups shared the common purposes of preserving emotional and pecuniary channels to their home villages as well as strengthening hometown ties among people in Hawai'i. The *kō* gatherings promoted fund-raising campaigns and carried out religious ceremonies, including special services for drowning victims. Hosting various recreational activities, such as athletic events, lotteries, and *bon* dances, was also part of *kō* support.[63]

Work and life patterns peculiar to the fishing communities produced specific psychological attitudes somewhat "different from those who were making their living on and from the land," said Kokuichi Yoshimura, secretary of the Fishermen's Association in Honolulu. Yoshimura said, "Those who engage in fishing do not get deeply involved in affairs of the land," and some of the society members showed little interest in or agreement with the activities of the Japanese society, which was under the domination of the "land people."[64] The distance between the people of the sea and those of the land was manifested in the indifference of fishermen to the Oʻahu sugar plantation strike of 1920, in which 8,300 Japanese and Filipinos, or about 77 percent of the plantation workforce on Oʻahu, walked off their jobs.[65] Among many letters from Hawaiʻi, which appeared in the monthly journal *Kamuro,* published in Okikamuro by the local youth society from 1914 to 1940, little mention was made of the strike, an earthshaking event for land people. The absence of the strike in the correspondence between Hawaiʻi and Okikamuro indicates that the fishermen and their families kept the foundation of their occupation and family lives without being heavily influenced by plantation incidents.[66]

Generational change and the problem of successors

Throughout the 1920s and 1930s, the Japanese fishing business enjoyed its golden age, blessed with an overwhelming number of sampans,[67] lively marketplaces under stable management, and magnificent shrines protecting people from the dangers of the sea. However, during the same period, a dark shadow slowly began to loom over the industry. One of the major problems that haunted Japanese fishermen was the paucity of successors. By 1929, the average age of Japanese fisherman was advancing because the nisei [second generation] did not follow in their veteran fathers' footsteps. Among the 736 fishermen affiliated with five Japanese fishing companies in Honolulu and Hilo, 654 were thirty years old and over, 82 were under thirty, and, most significant, only four were teenagers.[68] Usually, it took years of training for a young novice to become a full-fledged fisherman, so the extremely small population of young men alarmed the fishing establishment in Hawaiʻi. Without successors, fishing skills, an important manifestation of Japanese fishermen's ethnic identity and a means of affirming links to their home villages, were expected to disappear.

The decline of the younger population at sea can be interpreted as part of a larger trend in the Japanese community, the ever-increasing number of nisei who aspired to professional jobs or civil service employment in education, medicine, business, law, and government. Since Hawaiʻi in the 1920s and 1930s remained a predominantly plantation economy, the white-collar labor market was not broad enough to absorb the large numbers of educated nisei. And racial discrimination made this narrow gate even narrower. Reverend Takie Okumura, a Christian leader in the Japanese community, bemoaned the fate of the young, jobless nisei and urged them to reconsider plantation employment. Using the slogan "Back to the Land," he advised them to acquire professional knowledge and skills in agriculture, to buy land or homestead, and to live as yeoman farmers, in compliance with American traditions. The creation of agricultural schools to provide training in growing pineapple, raising cattle and poultry, and other forms of agriculture was also part of his agenda.[69] Okumura's argument was not, however, convincing enough to cause an exodus of nisei from the city to the plantations. To them, returning to the plantation and engaging in grueling labor under the haole (white) oligarchy was nothing more than accepting second-class citizenship.[70]

Fishing seems to have made a negative impression on younger generations, who viewed it as difficult and dangerous labor. From 1897 to 1931, thirty-two fishermen died or went missing during fishing operations. The powerful waves of Hawaiian waters, which often swallowed human lives, and the fear of storms and subsequent shipwrecks were ongoing concerns to those who ventured hundreds of miles out to sea. Even if their courage overcame their fears and they returned safely with rich hauls, fish prices did not seem high enough to offset the grave risks involved. The number of casualties (less than one fisherman per year perished at sea), seemed small compared to the various fatal accidents that occurred on land.[71] Nevertheless, a fisherman's life was always full of danger, as indicated by the widely known saying, *itago ichimai shita wa jigoku,* or, only a plank separates the fisherman from Davy Jones's locker.

> There is no occupation tougher than fishing, and there is not a more unrewarding, miserable life than a fisherman's. Although it is too late for us to change our jobs, we never want our children to succeed us in our work.[72]

These words, spoken by Kanzaburō Ageno, a fisherman on Moloka'i, represented the true voices of his colleagues. Ageno came to Hawai'i with his family in 1908 when he was nine years old, and witnessed how his fisherman father suffered during his pioneering days in Kahului, Maui. After receiving strict training from his father, Ageno became a professional fisherman. Out at sea, he eagerly conducted research on fisheries and regularly published his ideas and suggestions in Japanese newspapers for the betterment of the industry and fishermen's lifestyles. He argued that risking lives during operations and the seasonality and instability of a fisherman's income, together with its social position, made fishing a much less alluring occupation and encouraged children to seek other careers.[73]

As part of his efforts to resolve the negative aspects of fishing and pave the way for possible successors, Ageno stressed the importance of vocational education for the nisei. In particular, he urged issei parents to send their children to fishery schools in Japan. Since Hawai'i did not have such institutions, he put forth a plan to create a fishery school to teach navigation, meteorology, fishing methods, marine biology, and other relevant subjects.[74] Ageno was not alone in his demand for these educational programs. Alex Korol, manager of Hawaiian Tuna Packers, joined him, pointing out the limitations of the present Japanese-style apprenticeship. Korol argued that young men could no longer stand the hardship of onboard training and preferred jobs on land, even if they had a lower salary. "If he receives proper technical training, a fisherman will be able to double his income and improve his social status," Korol asserted.[75]

There was a movement to establish a fisheries training course at McKinley High School during the early 1930s. McKinley, called Tokyo High because of its large population of nisei students, was located in Honolulu near Kaka'ako. The plan clearly aimed at the recruitment of fishermen's children. Miles Cary, principal of McKinley, and directors of the territorial government's Department of Public Education and Division of Fish and Game joined the effort.[76] Cary usually rejected vocational education and focused more on academic majors, journalism, languages, mathematics, social sciences, and science.[77] His interest in fisheries education was, therefore, a remarkable exception. An English editorial in the Japanese newspaper *Nippu Jiji* also argued as early as 1929 that it would be "ridiculous" to expect young Japanese to be satisfied with work on the plantations under the present conditions; rather, it urged them to enter fishing, citing the great potential of Hawaiian waters. With adequate fisheries research and the improvement of efficiency in operations, *Nippu Jiji* wrote, Hawai'i could become one of the most lucrative fishing regions in the world.[78] *Nippu Jiji*'s owner, Yasutaro Sōga, who had played a primary role in establishing the oldest Japanese fishing company in Honolulu, looked to the sea as a place to accommodate a frustrated younger generation.

The demands for vocational training to replace traditional apprenticeship, the modernization of operations by adopting the latest fishing gear and methods, and extensive scientific research in an attempt to make fishing more attractive shaped the core of the "Back to the Sea" movement. It was a counterpart of Okumura's "Back to the Land" proposal to recruit as many nisei into plantation work as possible. Okumura's argument attracted support from haole plantation leaders because it had the potential to perpetuate Japanese economic dependency without overturning the foundation of the planter oligarchy. On the other hand, the "Back to the Sea" efforts, which aimed at preserving Japanese dominance at sea, gained only sporadic appeal and failed to earn enough influence to realize its goals. Throughout the 1930s, fisheries research remained poorly funded, and neither a vocational training course at McKinley High School nor a fisheries school ever came into being.

However, some of the nisei population responded to the enthusiasm of the issei pioneers. In 1936, Yoshio Hamamoto, a nisei fisherman affiliated with the Pacific Fishing Co., built a sampan that cost $7,500. He operated it himself as captain, with forty-five hired fishermen. Hamamoto was the first nisei to build a new ship, and his determination and dedication were good news for the aging veterans.[79] In sharp contrast, Tokiharu Takenaka, son of fisherman Isematsu Takenaka, from Wakayama, left his father's boat when his first child was born. He abandoned the sea at the request of his wife, who wanted him to spend more time with their baby. Thereafter, he started a welding business, and sometimes did ship welding, utilizing his knowledge of sampan structure. But he was less and less involved in fishing as time went by, and he never encouraged his son, Brooks, to become a fisherman.[80] As the Takenaka family episode reveals, the aging population of Japanese fishermen was not fully replaced by their children, who knew the difficulty of the occupation. In the meantime, the presence of US citizens of other ethnicities at sea slowly and steadily grew. According to statistics from around 1940, the number of citizen fishermen had surpassed that of noncitizens by 1940.[81] This record did not reflect the actual situation because some sampans had nominal citizen owners and left all the work in the hands of issei noncitizens.[82] Regardless of these cases, the aging of the Japanese population at sea continued throughout the 1930s.

The shrinking fish supply and new experiments

Surrounded by the vast Pacific Ocean, the aquatic resources of the Hawaiian Archipelago seemed inexhaustible, but the nonstop Japanese operations gave rise to increased apprehensions among some scientists about overfishing and the subsequent depletion of fish along the islands' shores.

As early as 1903, Oliver P. Jenkins, professor of physiology at Leland Stanford Junior University (now Stanford University), warned that "some of their [Japanese fishermen's] methods are very destructive, and if not regulated by opportune and wise legislation, will soon disastrously affect the fish fauna as a food supply."[83] The fear of overfishing and the depletion of aquatic resources was also felt among the Japanese. Torazuchi Hayashi, one of the founding members of the Suisan Co. in Hilo, regretted that overfishing emptied not only areas in and around Honolulu Harbor but also coastal waters off Oʻahu. According to Hayashi, these areas had been "swept with a broom" during the first decade of the twentieth century, and it would not be long before fishermen had to go to the island of Okino-Torishima, 1,087.5 miles from Tokyo, to fish.[84] His prediction was proven right within a couple of decades, when some fishermen left the depleted shores of the Hawaiian chain and headed to Midway. Even waters near the island of Okino-Torishima were considered potential fishing sites during the 1920s and 1930s because the growth in the size of sampans, their motorization, the enlargement of refrigeration equipment, and the improvement of fishing methods enabled such long-distance voyaging.[85]

The fishing communities in Hawaiʻi did not stand idly by and let aquatic resources disappear at home. While exploring new fishing grounds in distant waters, they attempted to increase stocks of marine products and add new types of aquaculture to the seascape of the Hawaiian Islands. The cliffs of Hawaiʻi's coastlines and its shore reefs precipitating sharply into the deep sea provided inhospitable conditions for rearing shellfish and seaweed, but the farming of *ina* (rock-boring sea urchin, *Echinometra mathaei* and *E. oblonga*), and *awa* (milkfish, *Chanos chanos*) in fishponds became more popular than other styles of aquaculture during the 1920s and 1930s. About the same number of Chinese and Japanese engaged in cultivating these fish, which provided them $20,000 to $30,000 worth of product to sell to fish markets every year.

The release of clams brought from Japan on the shores of Honolulu, Kalihi, Pearl Harbor, and ʻEwa was part of a variety of new experiments at that time. The clams spread rapidly. In 1925, Dr. Chiyomatu Ishikawa, from Japan, stocked the rivers of Kalihi, Kahana, Waimea, and Wahiawā on Oʻahu, and Waihina on Kauaʻi, with about 250,000 sweetfish eggs; he also introduced oysters from Virginia. The US government supported the release of tens of thousands of young trout into various rivers on Kauaʻi, Oʻahu, Hawaiʻi, and Maui. In the meantime, the territorial government imported 33,000 trout eggs and released them into rivers, and stocked the coast of north Oʻahu with hard clams from Japan. Transplantation of pearl oysters and Samoan crabs from other places was also a part of these attempts to create new Hawaiian fisheries. Some of the experiments seemed successful because released trout eggs hatched and grew enough to please anglers. However, the efforts did not lead to the significant development of new fisheries and the opening of new markets. What was worse, the overpropagated Samoan crabs soon started killing fish in fishponds and destroyed the delicate ecosystem of Hawaiian shores.[86]

4 Surviving the Dark Days

The increasing suspicion of the federal government and its escalating suppression of Japanese fishing

Japanese fishermen advanced into Hawaiian waters predominantly by their own will and efforts. Unlike the Japanese entrance into the seascapes around Korea, Taiwan, and other places, under the umbrella of the expanding Japanese empire, which gave fishermen various forms of official protection and support, their equivalents in Hawai'i could expect no direct pecuniary or material assistance from their home government. Instead, they used hometown ties and other forms of mutual aid to expand their fishing enterprises, as discussed previously.

Despite their autonomy from Japanese imperial control, the remarkable increase in the number of Japanese nationals at sea and their monopoly of the islands' fishing industries aroused suspicion among US federal authorities that they were acting as agents for the empire. The Hawaiian Labor Commission, formed by President Warren Harding, paid special attention to the extraordinary performance of Japanese fishing sampans capable of transporting thousands of men and sailing within a radius of 500 miles, and transferred this information to Labor Secretary James J. Davis. George Brooke, head of the Hawaiian Department's G-2 (military intelligence) section, also pointed out, in a report titled "Estimate of the Japanese Situation as It Affects the Territory of Hawai'i from the Military Point of View," that Japanese fishermen had accumulated substantial knowledge of reefs and harbors around the archipelago. For this reason, he suggested depriving all aliens of fishing rights in the waters of the

United States. Brooke's successor, Captain A. A. Wood Jr., went on to warn of the potential for Japanese fishermen, some of them having served in the Japanese Navy, to assist in a foreign power's attempt to capture the islands.[1]

The mounting apprehension among federal authorities and the military was manifested in fisheries-related regulations intended to disrupt the increase of foreign-born fishermen. In 1930, Washington gave a new interpretation to article 449 of the Customs Regulations, originally enacted in 1923, and ordered customs collectors to assess duties on all fish or marine products brought into the United States from the high seas by vessels owned by noncitizens. The expected tax rate was one to three cents per pound. This new order immediately caused serious confusion among those in the fishing industry in Hawai'i, where the Japanese owned and operated about 400 sampans.[2] While some harbored an optimistic view that it would produce a negligible impact, others felt it would deliver a death blow to the industry as a whole. The former optimists expected customs officials to be favorable to and protect the Japanese fishermen. Because the fishing industry had grown to be an enormous proportion of the islands' economy and an important provider of a protein staple for the local population, it seemed impractical to cripple the business by applying the strict new rules.[3]

Conversely, the pessimistic segment of the community thought that even the Japanese fisheries monopoly could not guarantee a minimum impact of the new order, for three primary reasons. First, the definition of "fish brought from the high seas" remained ambiguous. Sampans crisscrossed the high seas without paying much attention to dividing lines. If a boat passed through a regulated ocean area, caught fish in the unregulated territorial waters of other islands, and came back and sold the haul in a market in Honolulu, would it be taxable or not? Second, the definition of taxable fish was unclear and customs inspectors did not have detailed taxonomical knowledge of fish caught and consumed in Hawai'i. None of them knew whether skipjack tuna was categorized as a species similar to yellowfin and bluefin tuna. Third, a gap in the working hours of the customs office and the times of fish auctions was expected to cause trouble at the fish markets. Customs officials usually started work at 7:00 a.m. at the earliest, which was exactly when the fish bidding finished; many fishing boats unloaded their catch as early as 5:00 a.m., and fish markets kicked off their auctions at 6:00. If the daily round of business had to be adjusted to the regular working hours of the customs office, fish would sit for more than an hour before being auctioned. The freshness of fish was closely related to its value and price, and the delay would significantly lower its market value and jeopardize the livelihood of those in the fisheries business.[4]

When the new regulation was first enforced in May 1930, the optimistic hypothesis was demolished. Customs officials mercilessly began charging one to two cents per pound on marlin, swordfish, shark, and other fish. They showed up at auction sites at 5:00 a.m. and exempted skipjack, bluefin, and yellowfin tuna, the major products of the high seas, from taxation. However, they demanded that fishermen report to the customs office before going out to sea, which delayed their operating schedules. Fishing companies suffered, as well. The new rule increased the amount of paperwork, and they had to stop collecting commissions from fishermen who received only negligible profits after the deduction of taxes.[5] The enforcement of the new rule started in Honolulu, but was expected to spread to other parts of the Hawaiian chain.

As a predictable consequence, the islands' fishing industry started an aggressive and desperate campaign to abolish the regulations. Their agenda quickly went beyond the industry and became integrated into the strategies of the Japanese society and the Japanese Chamber of Commerce in Honolulu and Hilo. As these organizations pressed the territorial government and the Japanese consul general to take quick and efficient measures, they approached and won the support of Victor S. K. Houston, the only delegate from the territory of Hawai'i to the US Congress. Houston protested the regulation and had the US Department of the Treasury reexamine whether the pending rule conflicted with the treaty of commerce between the United States and Japan. Finally, he succeeded in having it repealed. The Japanese Chamber of Commerce immediately sent a telegram of thanks to Houston in Washington.[6] While Houston worked for the islands' commercial fishing interests, haole business leaders in Hawai'i, in particular, the white members of the Chamber of Commerce in Hilo, unhesitatingly joined the Japanese protest and publicly disagreed with the taxation on fish, mainly because consumers would suffer from increased fish prices.[7]

The collaborative efforts between the Japanese and white business leaders fended off interference from the federal government and protected the local fishing business, but they could not completely dispel the growing fear of Japanese domination of the islands' waters among federal authorities. Particularly, the US Navy viewed sampans navigating around Hawaiian waters with deep skepticism, fearing that they were spying for their mother country by supplying its submarines with information and transporting intelligence or sabotage agents to the islands. Their constant presence in Pearl Harbor and Kāneʻohe Bay amplified the Navy's anxiety because both had strategic naval bases. Ironically, both areas provided the best fishing grounds for baitfish on Oʻahu.

During the 1930s, skipjack tuna fishing reached its limit in productivity due to the insufficient supply of live baitfish. In sharp contrast to the abundant availability of skipjack tuna in open water, the poor catch of live baitfish crippled fleet operations and distressed cannery production, which depended on the local skipjack tuna supply. The Hawaiian Tuna Packers produced 110,000 boxes of tuna cans in 1930, an amount that plunged to 37,000 the following year. Seriously alarmed by the situation, the cannery transported ten tons of sardines from the mainland and released them in Hawaiian waters. But the average size of a fish from the mainland United States was larger than a local one and they immediately sank deep into the water, without attracting a school of skipjack tuna.[8] Despite this failure, the cannery repeatedly attempted to bring a supply of anchovies or sardines from the coasts of Mexico and Southern California throughout the 1930s. The Hawaiʻi territorial government supported these experiments. In 1935, at the request of Paul Beyer, manager of the Hawaiian Tuna Packers, H. L. Kelly, director of the Division of Fish and Game, made an inquiry to the Bureau of Commercial Fisheries of the California Fish and Game Commission concerning the transport of small fish over long distances. Later, Kelly approached the American Fishermen's Tug Boat Association in San Diego and asked about the suitable size of a ship and the time necessary to transport live baitfish.[9]

In addition to working with the territorial government, Beyer appealed directly to the US Navy to lift the rigid regulations on Pearl Harbor. Since the sampans were prohibited from entering certain harbor areas in and around the Navy yard and the entrances to Pearl Harbor, Beyer stated the following in a letter to Admiral H. R. Yarnell:

> This condition [scarcity of live bait] is a serious one, not only to the fishermen and their families, but to our cannery, which is dependent upon the fishermen for supplies of fish, as well as four or five hundred employees and their families who in turn are dependent upon the cannery for support.[10]

In order to save the fishermen of eighteen affiliated skipjack boats and protect the 400 or 500 employees of the cannery, he asserted that this distressing situation should be alleviated by making inner Pearl Harbor available to the industry as baiting grounds.[11]

> Because a can of tuna produced in Hawaii is welcomed in mainland U.S. for its good quality, its supply cannot meet the demands regardless of recent recession. If a large amount of skipjack tuna is available, the tuna canning industry has bright prospects.[12]

Throughout the 1930s, such a statement appeared in every edition of *Nippu Jiji*'s *Hawai nenkan,* an almanac of the Japanese in Hawai'i. Without increasing the access to the fishing grounds to acquire baitfish in Pearl Harbor, skipjack tuna fishing and the cannery would not have bright futures. Despite repeated requests from the cannery, the military did not loosen its tight control of Hawaiian bays.

Behind the suspicions of the military was Japan's resumed military aggression in China in the late 1930s. An unplanned clash between Japanese and Chinese troops in the Beijing area on July 7, 1937, quickly spread into a general war. Coincidentally, Japanese long-distance fishing fleets, consisting of trawlers and on-site factory ships, appeared in the high seas off Alaska for salmon fishing and acquired a notorious reputation as lawbreakers and predators engaging in heedless overfishing.[13] Although fishing fleets in Hawai'i were irrelevant to the activities in Alaska, the authorities' fear of Japanese residents escalated throughout the 1930s in Hawai'i and resulted in uprooting *mikkō-sha,* or those who came illegally from Japan and engaged in fishing. The smuggling had been a tacitly accepted, if not legitimate, form of transpacific movement among Japanese fishermen for decades. Nevertheless, the authorities suddenly started checking their backgrounds and expelling them from the archipelago one after another. As a result, many illegal fishermen had been deported back to Japan by 1940.[14] The exact number remains unknown because many of those who evaded deportation kept silent on the issue, even to their family members. Shizue Shimizu saw that at least four or five fishermen from Wakayama, all of whom came to Hawai'i at the invitation of her fisherman father, Isematsu Takenaka, went home "because US-Japan relations had become so thorny in the late 1930s."[15] Whether they left Hawai'i because of deportation orders or spontaneous decisions is unclear. They may have felt that the increasing pressure of various regulations made Hawai'i a less attractive fishing ground, or they may simply have thought about their age and realized it was time to go home and retire.

Tsurumatsu Kida was one of many fishermen who chose to leave Hawai'i in February 1941. His son, Katsukichi, explained that aggravated US-Japan relations compelled his father to hurry back to Japan.[16] But a new federal regulation issued in 1939 is a more convincing explanation for his sudden decision to leave Hawai'i, where he had devoted most of his life to the development of the fishing business since arriving from Wakayama in 1907. Since the late 1920s, noncitizen fishermen had obtained fishing licenses for $5 per year, while licenses for citizens were free. All fishermen, irrespective of citizenship, paid fees for the annual registration of boats, certain fishing gear, and nighttime operations.[17] The new rule of 1939, however, specified that only US citizens could have licenses to own fishing boats of five gross tons or more and engage in commercial fishing as captains. Most of the fishing boats operating in Hawaiian waters in those days were more than five gross tons, and far fewer citizens than noncitizens worked at sea. Despite the fact that strict enforcement of the new rule might expel Japanese fishermen from the sea, many of the Japanese sampan owners kept operating even after the new licensing law became effective in March 1940.

On February 28, 1941, a federal grand jury indicted seventy-one individuals, including Tsurumatsu Kida, and the "big three" Japanese fishing companies in Honolulu, namely, the Hawai'i Suisan Co., the Pacific Fishing Co., and the Honolulu Fishing Co., charging conspiracy in connection with the false licensing of sampans. According to US attorney Angus Taylor, the persons indicted took the countermeasure of ostensibly selling their licenses to US citizens by drawing up false bills of sale to their citizen relatives and friends; the fishing companies assisted the noncitizen owners in falsely registering their sampans to citizens. These citizens were, according to the charge, "poorly paid clerks, housewives and laborers not connected with the fishing industry."[18] In fact, the names of the indicted included several Japanese women, nisei sons, and influential figures, including Paul Beyer of Hawaiian Tuna Packers and Matsujirō Ōtani. Soon, US customs officials seized nineteen sampans, twelve of which were for skipjack tuna fishing and the rest for other kinds of tuna, and tied them up at Pier 16. Federal Judge Ingram M. Stainback set a blanket bond of $2,500 each. Later, the bond was reduced to $1,000, except for the executives of the three fishing companies. This action showed that the federal court targeted only the largest types of sampans in Honolulu. The wave of oppression was expected to extend beyond Honolulu.

Tsurumatsu Kida narrowly escaped the procedures. He left for Japan one month before the grand jury charged him, and he never returned to Hawai'i. He passed away in 1945 at his birthplace in Wakayama. However, other indicted men and women, including Tsurumatsu Kida's son, Katsukichi, a noncitizen fisherman, faced severe punishment. Katsukichi and other Japanese owners, who had suddenly lost their investments in boats and equipment, were forced to endure prolonged court struggles and bear considerable financial burdens. All of the accused took one of roughly three tactics: four pleaded not guilty, and the rest either pleaded guilty or, like Paul Beyer, Matsujirō Ōtani, and the other executives of the fishing companies, hired lawyers and submitted pleas of procedural hindrance.[19] Itonosuke Tamura, a fisherman from Hiroshima, took the first choice. He proclaimed his innocence against the charge that he had falsely registered his sampan to his two nisei sons. The federal court supported his argument and returned his boat, *Fuji-maru*.[20] In the meantime, Katsukichi Kida and the others who had pleaded guilty attempted to take back their sampans by hiring lawyers and petitioning the US departments of Justice and Commerce. Their actions bore some fruit: federal judge Stainback decided to return the sampans to their owners on the condition that they would sell them to American citizens endorsed by the court and pay fines of 20 percent of the value of the boats plus trial costs.[21] By October 1941, the chain of court battles was over, and the last tied-up sampan had been released.

The confiscation of the large sampans led to an estimated 20 to 30 percent cut in the daily fish supply. It was in the middle of tuna-fishing season, and the fishing companies assigned boatless crew members to the remaining vessels and stabilized fish prices as well as they could by increasing the amount of catch per boat.[22] Twelve of seventeen total skipjack tuna boats in Honolulu were taken away during the low fishing season, so the market price of skipjack tuna was relatively impervious to the decline in supply. Still, the prolonged tie-up of boats considerably disrupted the operations of the Hawaiian Tuna Packers, which needed large quantities of fresh skipjack tuna.

In addition to monetary and material losses, the Japanese fishermen had to fight against intensified suspicion of espionage. When the *Honolulu Advertiser* (the local English newspaper) reported, "the local fishing companies are controlled by the Japanese fish trust of Tokyo," it stirred angry sentiment among the Japanese.[23] The Japanese newspaper *Nippu Jiji* immediately sent its reporters to US attorneys and high-ranking customs officials and found that the fraud charge was unrelated to espionage. The *Nippu Jiji* took the further action of dragging the editor in chief of the *Honolulu Advertiser* to federal court; Judge Stainback sentenced him to sixty days of probation for contempt of court.[24]

Even though the federal judiciary clearly denied the connection between the sampans and Tokyo, the US Navy did not. By an executive order issued in February 1941, the Navy banned the Japanese fishing fleet from Kāneʻohe Bay, a rich fishing ground. This ban, issued for national security reasons, caused further distress to the local fisheries, which had suffered from oppression for more than a decade. The increasing limitation of fishing operations around naval bases led the FBI to charge Japanese with crossing the defensive zones at Pearl Harbor and Kāneʻohe Bay.

In contrast to escalating federal oppression, the local white business and political leaders sent favorable signs to the Japanese fishermen. A couple of days before the federal grand jury's indictment on February 28, 1941, the Honolulu Chamber of Commerce adopted a new resolution to promote local fisheries by lobbying a bill through the territorial legislature and Congress for the construction of a permanent fisheries research institution and a research ship. This new framework for activity was heavily influenced by "A Plan for the Development of the Hawaiian Fisheries," a report written by two experts from the US Bureau of Fisheries in 1939. This document stated that a fishery research vessel and permanent research facilities would be a prerequisite for the further expansion of commercial fishing in Hawaiʻi and demanded generous expenditures by both the territorial and federal governments to cover all the necessary costs.[25]

These ideas agreed with the agenda of the "Back to the Sea" movement in previous years, promoted by individuals who had encouraged the establishment of such facilities. The action plan of the Honolulu Chamber of Commerce meant that their long-term desire had finally been endorsed by the specialists at the US Bureau of Fisheries and in the white business community in Honolulu. Local politicians were also aware that fisheries research was one of the most inadequately financed types of research on food resources conducted in Hawaiʻi. In March 1941 when the local fishing communities were in the throes of the sampan seizure, territorial congressman Walter J. Macfarlane submitted a bill to provide $20,000 to stimulate fisheries. Macfarlane's bill covered the cost of exploring fishing grounds, studying fishing, considering job security and a reasonable profit for fishermen, and taking concrete measures to offer an adequate supply of fish at reasonable prices to consumers.[26] His vision was far narrower than that of the "Back to the Sea" contingent and the Honolulu Chamber of Commerce, but the bill was significant because the territorial legislature, which had given more weight to measures protecting aquatic resources and often conflicted with the interests of the industry, finally took up the matter of developing local fisheries for serious discussion. The favorable direction of local politics and business was, however, too weak and too late to change the more powerful current of history that was about to swallow up all the fishing boats in Hawaiian waters.

Pearl Harbor, the end of Japanese fishing

> On December 7, I [Yozō Masagatani] was fishing for *shibi* tuna off the coast of Waina [*sic*] in Honolulu and noticed a big, dense cloud of black smoke rising from Pearl Harbor.... At around 8:00 a.m., about thirty airplanes flew toward and signaled us by firing machine guns around our ships. Genshirō Sakashita said, "The war has begun. They were Japanese airplanes," but others replied, "We saw red crosses. They were informing us that something has happened on land." Around 10:00 a.m., American airplanes came. At the next moment, machine-gun fire crackled at masts and decks. We jumped into the engine rooms and fish tanks. They strafed us. After they were gone, we found an uncountable number of bullets stuck in the decks. Some said that they were real bullets, while others insisted that the strafing was nothing more than part of a military maneuver. It was so astonishing that we had no idea what we should do.[27]

Yozō Masagatani, a fisherman from Wakayama Prefecture, thus described the moment of Japan's attack on Pearl Harbor, which he witnessed off the coast from Honolulu on December 7, 1941. His statement conveys the great shock and confusion that he and his colleagues experienced at the unexpected encounter with Japanese airplanes and aerial attacks from US airplanes. Strafing navigating sampans was the culmination of suspicion by the US Navy and its ultimate form of harassment of the Japanese at sea. Although Masagatani and his colleagues survived the strafing, Sutematsu Kida, cousin of Tsurumatsu Kida, and his son, Kiichi, lost their lives when their boat was strafed by a US Army P-40. Suddenly widowed, his wife, Matsu, thereafter had to raise her six daughters by herself.[28] Several other Japanese fishermen and their families met the same fate.

The outbreak of war between the United States and Japan completely changed the seascapes of Hawai'i, from which all forms of fishing activity almost immediately disappeared. At 4:30 p.m. on December 7, martial law was declared. Governor Joseph B. Poindexter turned the territorial government over to General Walter Short. Soon, restrictions on commercial fishing boats navigating Hawaiian waters were specified in General Order No. 45 of the Office of the Military Governor. People of Japanese ancestry were prohibited from fishing because their presence at sea would pose a threat to national security. The US Navy impounded all the vessels belonging to Japanese resident aliens. Earnest Steiner of the Hawaiian Tuna Packers saw the Army or Coast Guard remove all Japanese sampans from the waterfront without thoroughly investigating each owner because "they had a lot of boats."[29] As the result of this drastic and indiscriminate measure, the number of licensed fishing boats ranging in length from twenty to eighty-six feet dropped from more than 500 in 1941 to about 130 in 1943.[30] Larger skipjack tuna boats were taken to Pearl Harbor and converted to armed inshore patrol crafts for the Navy because they were suitable for the open ocean, whereas smaller, dragline boats were hauled up the Ala Wai Canal.[31]

The confiscation of the sampans brought financial disaster to their Japanese owners. Kuniyoshi "James" Asari, an issei from Wakayama, lost his seventy-five-foot sampan, the *Tenjinmaru,* valued at $20,000. He had purchased it with all his savings and the support of friends, but the US Navy took it without any compensation. Furthermore, the Office of the Military Governor deprived him of his fishing license and forced him and many other veteran Japanese fishermen to give up their profession. They had no choice but to engage in land jobs, such as pineapple cannery work and defense-project jobs.[32]

Some fishermen lost their civil liberties and personal freedom in addition to their livelihoods for short or long periods of time. Unlike on the mainland West Coast, where approximately 120,000 Japanese residents and their descendants were indiscriminately sent to internment camps, Hawai'i imprisoned over 2,000 Japanese without clear evidence of a single incident of espionage or sabotage.[33] Most of them occupied leading positions in the Japanese community as Buddhist and Shinto priests, consular agents, language school officials, and *kibei,* who had spent their formative years studying in Japan. In Hawai'i, the Japanese and their descendants made up about 37.3 percent of the entire territorial population at the outbreak of war. Besides the impracticability of transferring a massive number of people to mainland camps by sea, where Japanese submarines were lurking, avoiding potential loss of labor and preserving the islands' economy were important reasons to forego immediate compliance with the order for wholesale incarceration.[34]

Shortly after the aerial attacks on his sampan on the day of the Pearl Harbor attack, Yozō Masagatani was approached by a US patrol boat, which escorted him back to Honolulu. When he and his colleagues arrived in port, they were all apprehended and taken away to the Immigration Office. Later, they were taken to the Sand Island military detention camp, across the bay from Honolulu. Masagatani remained interned for three weeks; after his release, he made his living at sawmills, earning about $1.20 per hour. It was, according to his recollection, far less than his earnings from fishing.[35]

Unlike Masagatani, who was freed after three weeks, Katsukichi Kida was sent to the mainland and held behind barbed wire throughout the war years. On the day of the Pearl Harbor attack, he was off the coast of Oʻahu. When he returned to Kewalo Basin, he was immediately arrested and taken to the Immigration Office, then to Sand Island, and, later, to mainland camps. At the time of his arrest, he had a wife and three small children; his youngest child, Donald, was only two years old. Donald later recalled how hard his mother, Yasue, a nisei from the island of Hawaiʻi, strove to protect the family and the family business during his father's absence. "My mother was running everything, running the store, and running the family," he said.[36]

The list of incarcerated Japanese included Matsujirō Ōtani. On December 4, three days before the beginning of the war, he celebrated the opening of Aʻala Market, which had just been restored from the ashes of a great fire that had swept through all the market buildings the previous year. As the new owner of Hawaiʻi's largest marketplace, with more than sixty tenants selling fresh and processed fish, vegetables, and many other items, Ōtani celebrated at the grand opening ceremonies and received blessings from many people from the fisheries as well as various other industries. His sense of accomplishment and resolution to embark on this new enterprise was soon replaced, however, by a feeling of tremendous confusion and despair.

In the early morning of December 7, he was busy preparing for what was to be a magnificent party, with hundreds of attendees; it would be the conclusion of a series of celebrations that had begun on December 4.

Matsujirō and Kane Ōtani in front of the A'ala Marketplace at its opening ceremony. When this picture was taken on December 7, 1941, Japan's attack on Pearl Harbor had already started. Photo courtesy of Akira Ōtani.

Inside A'ala Marketplace. Photo courtesy of Akira Ōtani.

Then he saw Japanese airplanes marked with the rising sun fly over the market and head to Pearl Harbor. People at the market, including Ōtani, were thrown into great confusion with the announcement that hostilities had just broken out at Pearl Harbor. Awhile after Ōtani went home, FBI agents and two soldiers marched into his house and dragged him into a car at gunpoint in front of his wife, Kane, and their children. Kane pleaded with them to wait a moment and let him put on a kimono and shoes. When her desperate appeal was bluntly rejected, she threw the shoes into the window of the car that was about to take her husband away. It was to be her farewell to him.[37]

Ōtani was taken to the Immigration Office, and then to Sand Island. On June 20, 1942, he and other internees were excited at the news that the authorities would allow them to see their families on the following day. Ōtani shaved his beard and prepared to meet Kane and his children. He waited all the next day, but his family never came. A new order reached him, sending him to a camp on the mainland. Deeply disappointed, he nevertheless accepted his fate as *shikata-ga-nai* (it cannot be helped). That night, he was put on a large steamship and taken to San Francisco. On his voyage across the Pacific, he joined Matsutarō Shimizu, a central figure in the Wakayama fishing community. He was also reunited with Katsukichi Kida and other fishermen friends in the mainland camps of Lordsburg and Santa Fe. Interaction with his old-time colleagues reduced Ōtani's sense of loneliness to a certain extent. Money, food, and clothes sent from Kane in Hawai'i, as well as the occasional visits of his sons, who were serving in the US Army, gave him the greatest comfort in the otherwise dreary life behind barbed wire.[38]

Whereas Matsujirō Ōtani and his family preserved a bond of affection through correspondence and visits, Matsutarō Shimizu lost communication with his adopted daughter, Shizue, who was stranded in Japan after the outbreak of the war. Shortly before the hostilities, his son, born in Hawai'i and not fluent in the Japanese language, began to suffer from a serious eye condition that the local doctors could not cure. Matsutarō asked Shizue to take him to a hospital in Osaka. Agreeing to his request, she went to Tanabe, where Matsutarō was born and his relatives resided, and accompanied the boy to an eye doctor in Osaka. Shizue remained until his eye disease was cured. Soon afterward, Shizue and the boy returned to Hawai'i. However, Shizue then went back to Tanabe alone because her relatives had asked her to help them. Then the war started. She had dual citizenship in both hostile nations, and her American citizenship made her life in Japan vulnerable to hostilities toward Americans among the Japanese people. In those days, strong anti-American sentiment pervaded Japanese society. Even some of her relatives treated her poorly. She remembered:

> I don't know why one of my aunts, a blood relative, always treated me as if I had been her enemy and constantly bullied me in a condescending manner. She did not know me well because I was born in Hawai'i.[39]

Besides emotional harassment, she was subject to economic hardship. Remittance from Hawai'i stopped, and her relatives refused to support her financially, even when devastating earthquakes and typhoons struck. She sold kimonos and a sewing machine that she had brought from home to cover the costs of fixing her house. Moreover, the suspicious eyes of her neighbors deeply bothered her. She heard some of the nisei residents in Tanabe slandered by charges of spying for the United States. Although she remained relatively free from direct accusations of spying, her dual citizenship gave the authorities an excuse to sometimes arbitrarily withhold rations of rice from her. Her struggle to obtain food continued after her marriage to Hisao Shimizu, her stepfather's relative. Soon they had two sons, but serious food shortages made her child rearing extremely difficult. She recalled the most painful moment with her child:

> When my second son was born, I did not have milk because of malnutrition. We had no rice. My son kept crying all night, pulling at my dry breast. My heart ached, because I let him go so hungry.[40]

Despite the chronic hunger and strong sense of loneliness, she never closed her heart to the people around her; when she had milk, she shared it with four or five babies whose mothers also suffered from malnutrition and could not breast-feed. She knew that food shortage was a common concern in the rural towns of Wakayama. When she extended a helping hand to others' starving babies, some of her relatives and neighbors reciprocated her kindness and supported her.

> Among my hostile relatives, a grandmother of my cousin was exceptional. She was always nice to me. Her husband, Uncle Hamaguchi, came to Hawai'i at the invitation of my natural father, Isematsu Takenaka. Both Uncle Hamaguchi and his wife knew me since my early childhood. They went back to Japan around 1937.[41]

The reunion with her relatives and old acquaintances from the Kaka'ako fishing community gave modest comfort to Shizue during the difficult years. Yet it was not powerful enough to obliterate all the bitter wartime memories ingrained in her heart. As recently as 2008, she said, "I don't want to talk about my experience in Japan, because it is never a good story. The Japanese don't want to hear it, and people in Hawai'i don't know it."[42]

Her reluctance to discuss those memories of Japan, with frequent reference to potential antipathy from a Japanese audience, reflects the negative treatment she received in Japan. Simultaneously, her statements indicate that the ordeal of nisei stranded in Japan has been absent from the collective memories of Hawai'i. The story has been silenced and marginalized by the dominant, masculinist narratives of the warring powers. The suppressed memories of Shizue Shimizu are, however, far from being isolated and uncommon wartime narratives. When the war started, more than 2,000 nisei were stranded in Japan, forcibly separated from their loved ones on the other side of the Pacific.[43] Shizue Shimizu was part of a segment of society that suffered from a transpacific family situation. It was not until after the war that she gradually became aware of the existence of other nisei residents in Japan, developed a sense of solidarity with them, and finally broke her silence to fight for justice.

Wartime negotiations to reestablish fishing operations

In 1942, the annual yield of deep-sea fish in Hawai'i dropped to 143,000 pounds, only about 1 percent of the yield of 13,428,000 pounds just before the war.[44] After its skipjack tuna fleet was requisitioned by the military, the Hawaiian Tuna Packers converted its plant to war work, offering its cold storage facilities for military use and its cannery production lines for airplane gas tanks. The Office of the Military Governor legitimized their policy as "precautionary measures in maintaining the defenses of the Hawaiian Islands." However, its officers had to confront the reality that the virtual elimination of "one of the most natural sources for food for the people of the Territory" might adversely affect the dietary habits of the islands' population.[45] Besides the large Japanese population, other ethnic groups in Hawai'i were willing to derive animal protein from fish. Importing marine products from the mainland and covering the deficit in the fresh fish supply, therefore, became a primary concern of the military authorities.

In Hawai'i under martial law, all of the commands of the military governor were issued as general orders covering the whole range of government affairs, and the Office of Food Control, one of its subdivisions, controlled the supply of necessary foodstuffs to incoming and outgoing military personnel and secured food for the islands' civilian population. The wartime trade between Hawai'i and the mainland was subject to many restrictions: the Office of Food Control prepared a monthly Master Priority List detailing the amount of certain items shipped from the mainland. This list included canned salmon, canned sardines, canned shellfish, dried fish, and other processed marine products, together with rice, wheat flour, meat, vegetables, and other foodstuffs consumed in Hawai'i. All local importers had to obtain permission from the director of the Office of Food Control before ordering and shipping commodities. But the selection of items being considered for importation was arbitrary, and the local importers were not adequately informed of when and how much they should purchase. They complained that the permit system was interrupting the smooth supply of truly necessary items to local consumers because they could not move the merchandise to the West Coast for storage.[46]

In September 1942, the Office of Food Control simplified the process of importation. Thereafter, importers only needed to file a carbon copy of each import order with the Honolulu Office of Food Control, although the Master Priority List continued to control the amount of certain imported items.[47] The change reduced the amount of paperwork for local importers to some degree, but it was not a panacea for the serious shortage of fish. Prior to the war, canned salmon was already popular among the islands' population and Hawai'i imported about thirty-four tons of it per month, besides consuming local fresh fish and locally canned tuna. The closing of Hawaiian Tuna Packers and precipitous drop in the local catch of fresh fish pushed up the estimated rate of canned salmon consumption to one hundred tons per month, according to H. H. Warner, assistant director of the Office of Food Control. Despite this, Hawai'i rarely received enough canned fish.[48] An increase in the amount of imported fish was of utmost concern to local importers, and some of them often gave "open orders for any kind of fish," which resulted in the importation of poor-quality, unsalable frozen fish.[49]

The Office of the Military Governor, including the Office of Food Control, were, as these haphazard actions indicate, staffed by amateurs in fish and fisheries, so they failed to develop efficient ideas after the termination of fishing activities. In the meantime, the local fisheries industry did not tacitly obey the arbitrary orders of the military or simply let their businesses die. Hawaiian Tuna Packers, in particular, assumed it was a matter of time before the military government would retract its regulations and allow local fishing to resume. To secure a leading role in shaping forthcoming strategies for sea production, Arthur H. Rice Jr., fishing superintendent of Hawaiian Tuna Packers, sent a letter to the Office of the Military Governor nine days after the Pearl Harbor attack. In his letter, Rice stressed the necessity of having a "competent agency" that would take complete charge of fishing operations and marketing arrangements for fish caught in Hawaiian waters, coordinate all matters with the military government, and administer fisheries-related regulations established by the authorities.

Rice, regarding his company as being "competent to act in this capacity," provided the latest information on local fishing. According to his analysis, Hawai'i had five types of fishing: trap, net, flag line, tuna, and small-boat hook and line. He recommended restarting the first two types of fishing in designated areas during daytime for two reasons. First, detailed information on trap and net fishing was available. Second, fishing boats for trap and net fishing could resume operations on short notice, with only Filipino, Hawaiian, and haole crews. Usually, these types of fishing required modestly sized boats and did not require mobilizing veteran fishermen of Japanese ancestry.[50] In order to identify and control the sampans at sea, Rice suggested painting them white, flying American flags from their mastheads, and painting American flags on each side and top of the cabins that could be easily seen from sea level and the air. By deliberating upon the tactical situation, putting the utmost priority on national security over the civilian use of the sea, his plan aimed at gradually rebuilding the fishing industry, beginning with the smallest scale of coastal fishing.

The Office of the Military Governor granted most of his requests. On December 30, 1941, the office appointed Hawaiian Tuna Packers as the supervisory agent of fishing and fish distribution, although its authority was limited to O'ahu. The next month, trap fishing and shore fishing was initiated off the coast of O'ahu, and some deep-sea boats were permitted to operate during the daytime only. Hawaiian Tuna Packers exclusively distributed their catch.[51]

In March 1942, Rice submitted a new plan challenging the current naval district order regulating fishing crafts: "No men of Japanese ancestry shall operate a fishing boat." Since there was an excellent flag-line ground for skipjack tuna off the Wai'anae coast of O'ahu, the plan suggested extending the fishing area from one mile out to sea under the current order to ten or twelve miles, and asked that fishermen of Japanese ancestry be included in skipjack tuna fishing. As a countermeasure against their alleged espionage, he worked out a plan of supervision. First, an escort boat, manned by a crew acceptable to Army and Navy authorities, could accompany this fishing fleet at all times and patrol around the boats during their operation. Second, on every trip, the entire fishing fleet would be cleared with the Coast Guard boat at Kewalo Basin upon leaving and returning to that port. Third, the fishing crews would stay aboard their boats when anchored off Waiʻanae, or, if they went ashore, would be checked in and out by the Army authorities there.[52]

Some of the requests to enlarge the approved fishing areas and expand the size of the fishing fleet were accepted, but the most important one, to tolerate a Japanese presence at sea, was rejected by the Office of the Military Governor. The military authorities remained unwilling to allow Japanese fishing, even after the victory of the Allied powers over Japan at the Battle of Midway in June 1942 completely expelled Japanese naval fleets from around the Hawaiian archipelago. One of the major projects of the military government was, however, to stimulate local food production, and the fishing skills of the Japanese were, as Rice emphasized, indispensable for expanding the local fish catch.

In November 1942, Walter F. Dillingham, newly appointed director of the Office of Food Production, removed Hawaiian Tuna Packers as the fishing coordinator, ostensibly to avoid the "awkwardness of regulating competitors' activities." He then installed Frank H. West of Hawaiian Cane Products in that post.[53] This decision surprised and confused West, who was an expert in the sugarcane industry but thoroughly unversed in the fishing business. West believed that his predecessor "had done a fine job," because "production [had] been as good or better than might reasonably be expected" under military restrictions and "prices had been kept [at] a very fair level." Immediately, West had a long talk with Alex Korol, manager of Hawaiian Tuna Packers, and asked for his opinions on the general problems of seafood production and marketing. The discussion convinced West that "Hawaiian Tuna Packers, Ltd. was the best local agency to handle the problem of deep sea fishing" and compelled him to write a letter to Dillingham urging Dillingham to arrange a tripartite meeting of Korol, the Office of the Military Governor, and the Navy.[54] Although no available records indicate whether Dillingham organized such a meeting, he agreed to West's proposal, and decided to set up separate agencies to handle offshore fishing and pond and shore fishing; Korol would coordinate the former and West would supervise the latter.[55] The power balance between the two positions was, however, asymmetrical, because West became the coordinator of the Fishing Division, a newly formed subdivision of the Office of Food Production, and Korol assumed the post of its assistant coordinator.[56]

The authority of these civilian coordinators covered the entire territory, but their attention was limited to Oʻahu, whereas the fisheries of the other islands had been largely placed under the military jurisdiction of district commanders. As the result of a consultation and agreement with the military authorities, West appointed Harold Frederick Rice as the deputy fishing coordinator for Central Maui, David Fleming for the West Maui or Lahaina area, Glen Mitchell for the eastern part of the island of Hawaiʻi, Leighton Hind for the western part of Hawaiʻi, and Alfred D. Hills for the island of Kauaʻi. F. W. Broadbent, deputy director of food production for Kauaʻi and former manager of Grove Farm Plantation, stated that they chose "some responsible haole" personnel who were familiar with the situation in designated areas. However, all of these deputy coordinators were ranchers and managers of an electric company and not directly involved with fishing-related business. The Japanese permeation of the industry was so widespread that it was difficult to find white men associated with fisheries, but those of Japanese ancestry were excluded as candidates.[57] The main function of the civilian deputy directors was to do what they could to increase the production of fish and to keep accurate records of production, so that they and the Honolulu office might be more closely coordinated. Giving licenses for fishing, wholesaling, and retailing marine products, accepting requests for gasoline allotments, and receiving reports on marketed fish were also areas of their responsibility.[58]

Through reinforcing the organization with civilian staff members and clarifying its duty and jurisdiction, the Fishing Division promoted the deregulation of complicated, compartmentalized military rules. Offshore fishing was, for instance, under the control of the Navy, and the Navy checked small boats doing onshore fishing as well as larger ones doing offshore fishing, even though onshore fishing was under the jurisdiction of the Army. The Army controlled anchorages for fishing boats within the fishing areas and checked the personnel and movements of these boats. This tangled allocation of authority hindered beach fishermen from using small, hand-propelled boats to stretch their nets. The military orders put the Army in control of fishing from the beach, but the Army local area commander could not approve such fishing activities using vessels, regardless of their size. West suggested changing the rule and giving local area commanders the authority to permit the use of rowboats for stretching nets from the beach. The Army granted his request.[59]

Thereafter, the Fishing Division kept pressing the military to allow more fishermen, Filipinos in particular, to expand designated fishing areas, and encourage pond fishing by allowing the construction of pond nets and traps. Besides acting as liaison between fishermen and the military, the division worked with the Navy and custodians of alien property toward the release of fishing boats taken over at the outbreak of the war. It also assisted fishermen in getting necessary clearance through the Navy and securing needed supplies and equipment.[60]

In spite of these efforts, as of January 1943, fresh fish landings were approximately 10 percent of prewar figures, despite a civilian population that had increased by 10–15 percent since the start of the war.[61] In those days, only twenty-one boats were engaged in deep-sea fishing and thirty boats and canoes were used to catch fish for pleasure and commercial purposes.[62] Without reestablishing large-scale offshore fishing, increasing the total catch of marine products and meeting the demands of local consumption was extremely difficult. Walter F. Dillingham, director of food production, stated, "a 100% restoration of pre-war fish production is not a prospect because of the necessary prohibitions against Japs [*sic*] using the high seas. Nevertheless a definite small increase may now be expected."[63]

David Fleming, deputy fishing coordinator of West Maui, wrote to West that the only way to ensure a more stable supply of fish was to resume skipjack tuna fishing:

> It is absolutely impossible to do such on Maui without a return to something akin to the methods formerly employed—use of proper boats, properly equipped and manned, and operating under conditions now incompatible with military orders.[64]

Maui had three skipjack tuna boats before the war, all of which were taken over by the military. The *Islander*, the best of the three, was operated by the Navy as a patrol boat. The wooden boats tied up in Honolulu had not been properly maintained and were probably not safe to put in deep water, and Fleming suggested that the Navy return the *Islander* to the service for which it was best fitted.[65] The *Islander*, equipped with a 200-horsepower engine, had been closely associated with the Japanese fishing industry on Maui. When Japanese fishermen, fish wholesalers, and dealers of Maui established the Islander Fishing Company in 1941, they purchased the *Islander* from the Hawaiian Tuna Packers and operated it as the mainstay of the company.

Fleming stressed the necessity of returning both the *Islander* and "at least a certain number of aliens" to Maui waters, questioning the necessity of distinguishing citizen and noncitizen fishermen. Certainly, he argued, the latter were "by far the better fishermen." If the authorities still suspected a Japanese presence in the open ocean, a Navy man should accompany the boat at all times and watch the operation. He also challenged the blackout restrictions prohibiting the showing of lights between the hours of sundown and sunrise as a major obstacle for bait gathering because fishermen caught *nehu* by using lamps suspended over large nets during the dark of night. Fleming stressed that Maui had rich fishing grounds for skipjack tuna. Other fish were caught and consumed on Maui, but they were seasonal and not able to match the abundant supply of skipjack tuna. His conversation with the former captain of the *Islander* convinced him that the *Islander* alone would be able to get five tons of skipjack tuna per day, twenty-five tons per week, or one hundred tons per month, and it would be a "tremendous help to the food situation on Maui where we are now down to enjoying (?) F. S. C. C. [Federal Surplus Commodities Corporation] meat."[66] It was "a shame" to set aside such a valuable food supply and depend on imported products instead. Imported shrimp was "all right for fishing off the rocks on a Sunday afternoon; but as a commercial proposition, it is a joke." He urged West to push Dillingham to convince the Navy to further deregulate fishing to increase local fish production.

In another letter to West, David Fleming conveyed the result of his meeting with a military commander. At the outset of the meeting, the commander adamantly opposed allowing "aliens or citizens of Japanese ancestry" on boats of any kind but gradually reversed himself and "offered to do his best to secure a relaxation of a number of regulations that now block *Aku* [skipjack tuna] fishing."[67] This change of attitude was a good sign for the revival of skipjack tuna fishing, but Fleming still felt uneasy about the commander's ignorance of the business in general. According to his observation, the commander did not seem to understand that skipjack tuna fishing was a specialist's job requiring specially trained people and expensive equipment, including the boat. Yet the commander believed that they "could get a supply of *Aku* from a little former *Ahi* [tuna] boat," although "*Ahi* boats were built for catching *Ahi* and other deep sea fish and are not at all fitting for catching *Aku*." As the dialogue between Fleming and the commander reveals, removing all the hindrances to highly productive deep-water fishing under martial law was difficult.

On October 24, 1944, martial law finally ended, but the end of military rule did not lead to an immediate termination of various restrictions on fishing. It was not until May 30, 1945, that Vice Admiral D. W. Bagley, US Navy commandant of the Fourteenth Naval District and commander of the Hawaiian sea frontier, issued an order to remove many of the restrictions on fishing in the territory. Thereafter, fishing areas and the times during which fishing boats could operate were substantially extended, and, most important, fishing licenses were granted to noncitizens. Nonetheless, lifting the restrictions imposed for three and a half years elicited a slow reaction from former Japanese fishermen and did not lead to a rapid expansion of the fishing fleet.[68] Heikichi Komine, a fisherman from Wakayama, looked back on those days:

> All of our boats were impounded by the Navy. After the war, we bought them back. . . . But these boats were seriously damaged. They still could be used, but they needed repairs because the Navy used them for four years.[69]

Komine, having started his fishing career as an illegal immigrant, rose up the ladder to be a sampan owner, and eventually became president of the Fisheries Cooperative Association in east Maui, when Japanese fishermen in that area formed it in 1941. Because of his position, the FBI apprehended him immediately after the outbreak of the war. When he was released eight months later, he found that his boat was gone. When his boat was finally returned after peace was restored, he was too tired of wartime prejudice and hardships, and probably too old to fix the battered boat and restart his business. Instead, Komine returned to his birthplace in Wakayama and stayed there for the rest of his life.

Military rule had gravely impacted the fish distribution system in Hawai'i as well as production. The crippled supply of fresh fish and the price controls of the Office of the Military Governor ended most of the jobs at fish auctions. Before the war, Honolulu had eleven fishing companies, but all of them disappeared during the hostilities or within two years afterward. The severely reduced amount of marine products caused great difficulty for fish wholesalers and retailers, but some managed to survive in the unfavorable circumstances. When the Office of Food Production introduced the import license system to replace the previous permit system in September 1942, it received application forms from M. Ōtani Company, Ltd. During the absence of Matsujirō Ōtani, its president, his children tried their best to protect their family business by handling products imported from the mainland.[70]

On the island of Hawai'i, the military took over the Suisan Co. building in Hilo and used its refrigerated storage for military purposes. Kamezō Matsuno, president of Suisan Co., was interned at the Kīlauea Military Camp during the war. His son, Rex, said, "The situation was better on the island of Hawai'i than in Honolulu because he was not sent to the mainland as was Matsujirō Ōtani." The loss of its president, facilities, and fish supply forced Suisan Co. to cease operations, but it soon opened an auction site in another place and restarted handling drastically diminished amounts of fish brought in by non-Japanese fishermen, mostly Hawaiian and Portuguese, who engaged in small-scale shore fishing. Rex Matsuno remembered:

> There was one fellow in particular named Manalili who used to transport fish from as far away as Miloli'i [an isolated fishing village on the southwest coast of the island]. If not for people like him, the fishermen and loyal customers, Suisan wouldn't have survived.[71]

The Suisan Co. somehow managed to survive with a skeleton crew until the dark days were over.

The war in the Pacific, which ended in August 1945, left deep scars among fishing households and communities. When Katsukichi Kida looked back at the history of skipjack tuna fishing in Hawai'i, he described the war years as follows:

> World War II was one of the most crippling factors in the decline of the tuna-fishing industry among the Japanese in Hawai'i. The disappearance of the fishing community of Kaka'ako is one evidence of this.[72]

The elimination of the fishing fleet in Kewalo Basin caused the destruction of the adjoining Kaka'ako fishing community. Because most of its residents lost their fishery-related work and took defense-project jobs or other land-based employment, they no longer needed to live in a particular space in which people of similar occupations nurtured solidarity and were mutually supported. Outside of Kaka'ako, people who lived too close to the water's edge were forced to move elsewhere for security reasons, which inevitably reduced the populations of fishing villages, most of which had developed along the coastlines.

The rehabilitation of Hawai'i's near-defunct fishing industry and restoration of the dominance of the Japanese at sea seemed very difficult. "It was like a dream in the old days.... You could look at the ocean from the boat and just see the tuna swimming around, waiting to be caught," Katsukichi Kida said of the golden days of Japanese fishing in Hawai'i.[73] Kida spent four years behind barbed wire at internment camps on the mainland, waiting to go back to Hawai'i and realize his dream one more time.

5 The Reconstruction and Revitalization of Fisheries after World War II

The revitalization of the fishing industry

> When my father came back from internment, we went to Pier 35. When he arrived there, my mother said, "This is your daddy," pushing me toward him. He didn't recognize me. My [older] sister started crying.[1]

This was the moment when Donald Kida greeted his father, Katsukichi, who had come back to Hawaiʻi from an internment camp. At the time of the Pearl Harbor attack, Donald was two years old. He reached school age without a memory of his father's face; instead, the image of a father in school textbooks, a Caucasian white-collar man in a business suit, substituted for his real one. Katsukichi, the dark and tired-looking man standing in front of him at Pier 35, completely betrayed his preconception. "My father was a stranger to me," he said. This initial impression at the moment of reunion lingered in his heart for years and hindered the growth of mutual affection between him and his father.

As Donald Kida's experience indicates, recovering what had been destroyed by the war was far from an easy process and was frequently accompanied by numerous changes and sacrifices. Katsukichi Kida wanted to go back to sea and resume fishing, but his wife, Yasue, strongly opposed: "If you stay on the boat and go fishing, I'm going to close the store." During the war, she single-handedly ran the general store Katsukichi had inherited from his father while also caring for the family; it was time for her to be relieved of the heavy burden of being the sole manager of the business. Because of the years of toil and hardship she endured during his absence, he gave up his old profession and turned his energy to managing the store, although he later changed it to a fishing supply shop, using his skills and knowledge of fishing.[2] As the war dramatically changed the seascapes of Hawai'i, so did it change the family lives and lifestyles of many fishermen; Katsukichi Kida's was one such case.

Unlike Kida, Kuniyoshi "James" Asari from Wakayama immediately went back to fishing because he assumed that the wartime prohibition on fishing had replenished many fish stocks and increased their density. Moreover, the end of the war unleashed a craving for fresh fish among the local population. The older but still reliable boats were sitting unused and waiting, though many large, well-equipped vessels, including his own, were gone or had been returned damaged.[3] Many other fishermen came back from their wartime losses of equipment and suspension of fishing operations, and returned to the fishing grounds one after another. Hawaiian Tuna Packers, which operated as a war plant during the war, swung back into tuna packing with a restored skipjack tuna supply. The recovery of fishing operations stimulated the appearance of small *kamaboko* and other fishery-related plants.

The reconstruction of fishing companies took a thornier and more complicated path. When Matsujirō Ōtani returned to Hawai'i from the mainland camps, he saw that none of the fishing companies on O'ahu had survived the turmoil of war and its aftermath. In 1947, he organized the Kyodo Gyogyo, a sort of fishing cooperative with former executives of defunct fishing companies, opened a space for auctions at his A'ala Market, and located its office on the second floor.[4]

As the only fish auction on O'ahu, the Kyodo Gyogyo soon fell prey to political rivalry because its owner, M. Ōtani Co., Ltd., was headed by territorial senator William H. Heen. Heen, a longtime friend and business partner of Matsujirō Ōtani, was one of the most powerful figures in the Democratic Party of Hawai'i throughout the first half of the twentieth century. In 1951, the allegation that its auction was a monopoly led to a bill introduced in the territorial legislature attempting to abolish the single auction by authorizing a fishermen's cooperative. The bill was, however, killed by a Senate committee. A new company, King Fishing Co., Ltd, opened on Kekaulike Street soon afterward, in 1951, and ended the Kyodo "monopoly" on fish auctions; Senator Hiram L. Fong, a Republican rival of Heen, became its director. The birth of a new firm caused

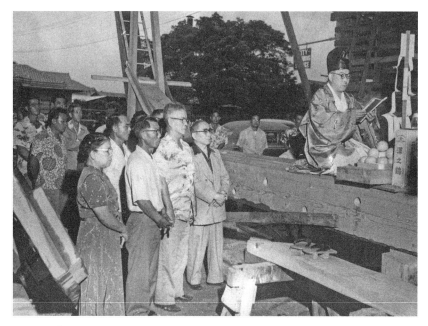

Matsujirō Ōtani, standing beside part of a boat that was being built, was the patron of many fishermen. A Shinto priest is reciting a prayer for a new skipjack tuna fishing boat. Photo courtesy of Akira Ōtani.

factionalism among the associates of the Kyodo Gyogyo and escalated into its dissolution. In 1952, Ōtani organized the United Fishing Agency, Ltd. Until the King Fishing Co. went out of business in 1968, Oʻahu had two auction sites run by these two companies, although the former had a larger number of associated fishing boats.[5]

Other parts of the Hawaiian chain also experienced ups and downs while rebuilding the fishing industry. In Hilo on the island of Hawaiʻi, a powerful tsunami in 1946 swept through the reviving Japanese fishing fleet, homes, and business, claiming many lives in the Hilo-Waiakea area and forcing many old-timers and younger people to give up fishing. When Suisan Co., the only surviving fishing company in Hilo, endeavored to reconstruct the infrastructure of the industry, many young people returning to the community from military service joined the company. Rex Matsuno, son of the company's president, Kamezō, urged diversification of the business and started frozen food operations. His new enterprise was innovative and ahead of its time because in those days, "nobody had freezers."[6]

The Reconstruction and Revitalization of Fisheries 119

At the launching ceremony for a fishing boat, people celebrated by scattering and picking up *mochi* rice cakes. Photo courtesy of Teruo Funai.

However, its frozen foods department soon grew to become the principal moneymaker, rather than the marketing of fresh fish.

During the late 1940s, the number of fishing boats in the Hawaiian archipelago significantly expanded; added to boats of prewar vintage, new ones equipped with the latest radios and fisheries sonar were constructed. The expansion of the fishing fleets accompanied that of the fishing population.[7] The vast majority of fishermen were still issei and nisei, but those of other ethnic backgrounds, in particular Native Hawaiians and Filipinos, both of whom played leading roles in protecting the industry from total collapse during the war, became integrated into the core of the fishing community. The postwar boom of commercial fishing impelled the revival of fishermen's organizations. In January 1952, younger tuna fishermen, most of whom were nisei, reorganized the Honolulu Tuna Fishermen's Association with approximately ninety members. This association, formed in 1936, was inactive during the war. In February of the same year, the Skipjack Tuna Fishermen's Association celebrated its revival from years of dormancy at a Japanese restaurant in Honolulu.[8]

Rebuilding ties with Japan

Subsequent to the acceptance of the Potsdam Declaration in August 1945, Japan came under the occupation of the General Headquarters, Supreme Commander of Allied Powers (GHQ/SCAP), led by General Douglas MacArthur (1880–1964). Matsujirō Ōtani tried to strengthen the relationship with Japan through the restoration of personal ties with influential figures in Japanese fisheries. In particular, he deepened the association with Ken'ichi Nakabe and his family during the occupation of Japan (1945–1952).

Ken'ichi Nakabe was the president of Taiyo Gyogyo, one of the leading fisheries conglomerates in Japan. In the 1930s, Hayashikane Shōten, its forerunner, started a partnership with Ōtani, exporting frozen tuna and other foods to Hawai'i. During the war, the destruction of Taiyo Gyogyo's physical resources, the depletion of the workforce, and the great loss of major fishing areas following the breakup of the Japanese empire tormented Ken'ichi Nakabe. What was worse, he was diagnosed with stomach cancer soon after the war. Since the medical institutions in Japan were dilapidated and medicines were in short supply, Matsujirō Ōtani invited him to Hawai'i.[9] Soon, Ken'ichi Nakabe flew to Honolulu, stayed at Ōtani's home, and received treatment at Straub Hospital. Years later, Ōtani hosted the scions of the Nakabes to study in the continental United States, including Tōjiro, who later became the president of Taiyo Gyogyo.[10] The deepening friendship between Ken'ichi Nakabe and Matsujirō Ōtani was a prelude to restoring the strong ties between Japan and Hawai'i at sea.

On February 6, 1953, the *Shunkotsu-maru,* a training ship of the National Fisheries University in Shimonoseki, arrived at Pier 7 in Honolulu Harbor. About three hundred people, mostly of Japanese ancestry, greeted its eighty-seven cadets, officers, and crew. The appearance of the vessel before the people of Honolulu symbolized the restoration of Japan's sovereign status with the San Francisco Peace Treaty, concluded on September 8, 1951, and effective April 28, 1952, and the subsequent reestablishment of transpacific ties. The local people, the issei in particular, heartily welcomed the crew of the *Shunkotsu-maru* and packed their schedule with public and private events. Some of them enjoyed reunions with their relatives, and others invited those from the same native places to lunch and dinner. Hawaiian Tuna Packers, the Japanese Club in Waipahu, the Japanese Chamber of Commerce and Industry, and the Honolulu Consul General held welcome parties attended by hundreds of issei to meet the

Ken'ichi Nakabe (at the right side of the second row) at Honolulu International Airport. Photo courtesy of Akira Ōtani.

Cadets of the National Fisheries University at a sumo tournament. Many people, those of Japanese ancestry in particular, held special sumo tournaments to welcome cadets, officers, and crew members. Photo courtesy of Yōzaburō Hayata.

guests from Japan. At a special sumo tournament at Kotohira Shrine in Kapalama, in which the cadets participated, a local Japanese newspaper reported that about 2,000 spectators "greeted student sumo wrestlers with thunderous applause."[11] The warm welcome was repeated in Hilo when the *Shunkotsu-maru* sailed there. The Japanese community of Hilo welcomed it with magnificent parties and a sumo tournament at Hilo Daijingu, a Shinto shrine built in 1898.

The fervent welcome extended by the Japanese communities of Honolulu and Hilo to the *Shunkotsu-maru* was an honest expression of their emotional attachment to Japan. Hideo Tagawa, a cadet on the *Shunkotsu-maru*, remembered the final moment of departure from Hilo:

> When the *Shunkotsu-maru* was about to leave, an old issei woman came on board to greet the national flag. Then, she started begging us with tears to take her to Japan. When we told her again and again that we would be back soon, she reluctantly left the ship. Because of that, our departure time was delayed. It was a really heartbreaking moment to sail for Japan, leaving her behind.[12]

This episode reveals how deep the nostalgia for their birthplace was in the hearts of issei Japanese. This feeling was hard to express during the war, when the Japanese community in Hawai'i was under great pressure to eradicate things Japanese. During the war, women stopped wearing kimonos on the street. Japanese national flags, swords, and any treasures from Japan were destroyed by their owners because they feared such nationalistic artifacts would link them to the enemy. Japanese-language schools closed, and people refrained from speaking Japanese in public. Buddhist temples and Shinto shrines, the priests of which were rounded up and sent to mainland camps, stopped services. While disassociating themselves from anything Japanese, the Japanese population in Hawai'i aggressively complied with the US war effort by promoting Americanism in all aspects of their daily lives. The young nisei expressed their loyalty to the United States by joining the military or the Varsity Victory Volunteers. In fact, the 442nd Combat Team, consisting mainly of nisei inductees, fought courageously and at great sacrifice in France and Germany, earning the title "Army's most Decorated Unit."[13]

Such great cultural suppression seemed permanent, but the end of the war reduced the pressure for Americanization and invited the rapid revival of Japanese culture and institutions. Soon the Japanese Chamber of Commerce and Industry of the issei was reorganized, Japanese movie theaters and restaurants were flooded with customers, and Japanese songs gained popularity among both issei and nisei Japanese.[14] This historical context produced the various Japanese organizations that sponsored the welcome parties for the *Shunkotsu-maru* and dozens of young local sumo wrestlers who faced the cadets at the tournaments at Kotohira Shrine and Hilo Daijingu.

The reestablishment of Shinto shrines from the wartime suspension of activities was fraught with difficulties. The federal government, suspecting that Shinto was the spiritual backbone of Japanese imperialism, confiscated the land and buildings of Kotohira Shrine as enemy assets and attempted to put them up for auction in June 1948. The patrons of Kotohira Shrine, many of whom were fishermen and their families, filed suit against US Attorney General Tom C. Clark, refuting the allegations with the argument that the shrine had functioned genuinely as a local religious institution without succumbing to control from Japan or aiding any anti-American activities. The rapid revival and expansion of the fishing fleet during the late 1940s regenerated the spiritual ties of its members to the shrine and enabled them to finance years of costly court struggles. Thanks to their dedication, the Kotohira Shrine regained all its property in 1951. Hilo Daijingu and other Shinto shrines in the archipelago faced a similar fate, but they followed the path of Kotohira Shrine and regained lost lands and buildings after filing and winning suits.[15]

The restoration of citizenship and family life for Shizue Shimizu

Among the boats of the reviving Japanese fishing fleet was the *Kinan-maru,* a newly built tuna longline fishing boat, fifty-eight feet long and fifteen feet wide, equipped with the latest Caterpillar engine. The owner of the boat was Matsutarō Shimizu. No sooner had he come back from the mainland camps than he resumed fishing under the patronage of Matsujirō Ōtani. Feeling deeply indebted to Shimizu, who had devotedly nursed Ōtani when he suffered from heart trouble at the camp, Ōtani financially and emotionally supported Shimizu's postwar endeavor. At the launching ceremony of the *Kinan-maru* in July 1953, Ōtani was included in a family photo in which Shimizu, his wife, Haru, his two stepsons, ship carpenter Seiichi Funai, and the owner of a shipyard were standing proudly in front of the new boat, which was decorated with a big lei at the head and flags. The photo, however, lacked one more indispensable Shimizu family member: Shizue, Matsujirō's adopted daughter. At the time of the grand event, she was still stranded in Japan.

The end of the war between the United States and Japan did not reduce the agony of Shizue's life in Japan. In addition to chronic food shortages, tremendous pressure to mobilize all eligible voters to participate in the first general election in 1947 brought her to a critical turning point in her life. This election has been commonly depicted as a pinnacle of democratic reform under the rule of the GHQ/SCAP; for the first time in Japanese history, women had suffrage, and thirty-seven women were elected to the National Diet. But the election, which symbolized the liberation of women from semifeudalistic shackles, had a completely different meaning to Shizue, a nisei woman: "If you don't vote, we don't give you any ration of food."[16] This threat of life-or-death reprisal from the authorities forced her to reluctantly participate in the election, and, as a result, she lost her US citizenship. Thereafter, she was determined to forget Hawai'i and continued to live in Japan with her husband and two sons.

Matsutarō felt responsible for Shizue's suffering because his request for her to fulfill a family responsibility had resulted in her remaining in

At the launching ceremony for the *Kinan-maru*. Matsujirō Ōtani, second from left; Matsutarō Shimizu, fourth from right; Haru, far right. Photo courtesy of Teruo Funai.

Japan. Once Japan's sovereign status was restored, he tried to bring her back to Hawai'i. Moreover, Matsutarō harbored a special passion for justice, not only for Shizue, but for all stranded nisei. Beginning in 1952, two attorneys at law, Katsurō Miho, who had just returned from Japan, and A. L. Wirin, started preparing to file lawsuits in the federal court of Honolulu on behalf of the nisei who had been unable to return to the United States from Japan during the war and had lost their US citizenship as a result of serving in the Japanese military, working for the Japanese government during the war, or voting in the national election. As an attorney for the Japanese American Citizen League, Wirin devoted his career to the civil liberties of nisei, including Gordon Hirabayashi, Minoru Yasui, and Fred Korematsu, all of whom challenged the forced removal and detention of West Coast Japanese Americans. Wirin also represented the nisei who had renounced their American citizenship at the Tule Lake Segregation Center and succeeded in restoring it to them.[17]

When Matsutarō Shimizu learned that Wirin and his allies were planning to come to Hawai'i and work for the nisei stranded in Japan, he wrote to Shizue:

> You are not alone in this fight. There are many other nisei who will fight for the restoration of citizenship. You are fluent in English. You should stand up in the court and explain what happened. I will pay all of the costs.[18]

Initially, Shizue hesitated to return to Hawai'i for trial because she was afraid that her Japan-born sons would be harassed in Hawai'i as "foreigners," as she was treated in Japan. But Matsutarō's passion finally overcame this fear and convinced her to join the class action together with other nisei. In 1953, she came back to Hawai'i to fight for justice, leaving her husband and sons in Wakayama.

"At first, I could not speak English fluently in the court, because I had lived in Japan for a long time," she said. Nevertheless, Shizue tried her best to speak English on the witness stand and testified about life in Japan, explaining how she was treated during the war, what kind of food she ate, how food rationing functioned, and so on. The trial lasted for about one year, and, finally, in November 1953, Federal Judge Jon Wiig ruled in her favor; Shizue's citizenship was restored along with that of eight other nisei plaintiffs. Matsutarō paid all the costs, which amounted to approximately $1,000.[19]

A decline in the population of fishermen and negotiation with the Japanese government

In 1955, Hisao Shimizu left Wakayama for Honolulu with his two sons to join his wife, Shizue. "I felt sorry for my little boys, whom nobody took care of at home while I was away at work, so I decided to come to Hawai'i to let them join their mother."[20] This was how Hisao explained why he sailed across the Pacific for the first time in his life. Upon arriving in Hawai'i, Hisao got onboard the *Kinan-maru* under pressure from Matsutarō. He suffered from serious seasickness out at sea because he had no previous fishing experience. However, Matsutarō had no alternative but to rely on Hisao for help in spite of his physical unsuitability for the job. By the mid-1950s, the postwar boom in commercial fishing, characterized by an increasing number of boats and crew members, was over, and the fishing population had sharply declined. During the 1950s, the number of commercial fishermen plummeted from 2,478 to 1,022, a loss of 59 percent.[21] Matsujirō Ōtani, of the United Fishing Agency, said:

> Small fishing boats had been replaced by larger ones, but no fishermen would go to the sea. Many boats were always floating at fishing ports, and many boat owners were on the verge of bankruptcy. Fishing companies faced a grave crisis. . . . I think the government has been too indifferent to the fishing industry vis-à-vis agriculture, although the former has been more important than the latter in Hawai'i.[22]

His statement was partially correct because both the federal and the territorial governments had remained indifferent to the fisheries for a long time. However, after the war, the federal and Hawai'i territorial (after 1959, state) governments belatedly started spending "a large sum of money" for the protection, administration, research, and development of the fisheries of the central Pacific and Hawai'i; the total amount had reached approximately $10–15 million by 1965.[23]

The postwar influx of public money into the fishing business was the result of officials' concern for the weakening industry or their reaction against the mounting rivalry with foreign fishing fleets advancing into the western Pacific. The governmental efforts, however, did not change the declining trend of the islands' commercial fishing, which suffered from two problems. First, the cost of fishing operations and other aspects rose markedly. Second, employment opportunities in other occupations for men with fishing skills became increasingly abundant.[24] The improvement of efficiency and technology could solve the first problem, but the second was far more difficult and complicated. As examined previously, the industry had been suffering from an aging and shrinking fishing population since the 1930s. The postwar boom temporarily solved this problem, but the earthshaking changes in the socioeconomic structure of Hawai'i produced numerous new job opportunities and made it hard to entice local boys into the fisheries regardless of the good wages.

Despite these problems, the local consumption of fish per capita in the 1950s was almost triple that of the mainland, and the demand for fishery products kept growing with the expansion of the islands' population.[25] In order to maintain a balance of supply and demand, Matsujirō Ōtani negotiated with Taiyo Gyogyo in Japan, led by the Nakabe family, and the *Banshū-maru*, its refrigerated cargo ship, arrived at Pier 10 in Honolulu on October 19, 1952, carrying a full load of six hundred tons of tuna for processing at Hawaiian Tuna Packers.[26] In the following year, the *Banshū-maru* returned to Honolulu with six hundred tons of tuna for the cannery and twenty-five tons of various fish for Ōtani's company. Thereafter, the amount of imported fish and fish products from Japan and other Pacific countries grew and supplemented the deficiency of locally supplied fresh fish.

Matsujirō Ōtani saw that shipping fish from Japan via the *Banshu-maru* could temporarily relieve, but not fundamentally solve, the problems of the industry. Since the early 1950s, he had visited Japan three times and repeatedly talked with the Ministry of Foreign Affairs about the possibility of recruiting Japanese fishermen. Senator William H. Heen accompanied him as a senior adviser of the United Fishing Agency and a politician representing the territory. Their joint efforts to persuade the Japanese government proved fruitless; their request was turned down flat.[27] During the 1950s, Japan was fully occupied with increasing domestic food production and considered fishing indispensable both as a pillar of the national economy and for the health of the people; fish provided vital animal protein to the still-hungry population. When Japanese fishing fleets were busy exploiting aquatic resources along the Japanese coastline, offshore, and in all major fishing areas of the world, with strong support from their government,[28] sparing fishermen for Hawai'i could be seriously detrimental to the national interest. Moreover, the nightmare of Japanese exclusion movements triggered by the flood of immigrants into the United States during the first decades of the twentieth century deterred the Japanese government from sending people there and creating further tension. These economic and political factors undermined Ōtani's plan.

While Ōtani was negotiating with the reluctant Japanese government, Okinawa was slowly but surely looming larger. In 1953, four fishermen from Okinawa arrived in Honolulu for training purposes. Their visit formed a preface to the next stage of fisheries history in Hawai'i.

6 Okinawa and Hawai'i

The historical background of commercial fishing in Okinawa

The Okinawa archipelago, stretching about eight hundred miles in an arc from the south of Kyushu, a southern island of Japan, to Taiwan, is an integral, albeit unique part of Japan with distinctive historical and cultural characteristics. When Okinawa, the principal island of the Ryukyus, was unified by Shō Hashi in 1422, Shō and the successive rulers of the Ryukyu Kingdom (the First Shō Dynasty, 1422–1469, and the Second Shō Dynasty, 1470–1879) paid tribute to the emperors of China and promoted foreign trade via ships to China, Japan, Korea, and Southeast Asian countries by fully taking advantage of its location, connecting East Asia and the South Seas. After Satsuma, a feudal domain in southern Kyushu, conquered the Ryukyus with the approval of the Tokugawa shogunate and made it a vassal in 1609, the Ryukyus were integrated into the sphere of Tokugawa Japan while preserving a facade of independence and sending tribute missions to China to continue trading.

Throughout the era of the Ryukyu Kingdom, agriculture was the chief means of production and was fundamental to domestic administration, whereas fisheries remained peripheral to agriculture and occupied a subordinate position in the islands' economy and politics. Although its active entrepôt trade had shaped the identity of Ryukyu as a great maritime nation communicating with many countries, the presence of fishing boats had always been overshadowed by that of trade ships. In the agrocentric history of the Ryukyus, Itoman, a village located in the southern part of the main island, was a distinctive center of commercial fishing. The dearth of arable land forced the people of Itoman to depend on fishing for their livelihoods, and the kingdom granted them various fishing rights in exchange for offering fish to the king and high officials. By producing the Ryukyus' major exports to China, dried shark fins, dried sea cucumbers, dried squid, seashells, and other precious marine products, Itoman supported the economic foundation of the kingdom.

After the Meiji Restoration in 1868, and the subsequent reorganization of the kingdom into Okinawa Prefecture, Japan terminated its tributary trade with China, and the demand for dried shark fins and sea cucumbers shrank. Instead, the fishermen of Itoman started gathering spiral shells to export to Europe as materials for luxury buttons. The invention of water goggles in 1884 enabled fishermen to engage in longer and deeper diving without damaging their eyes. Soon, they invented a unique fishing style in which approximately thirty to fifty fishermen dove into the water from five to ten small, hand-paddled canoes or *sabani*, swam, and drove

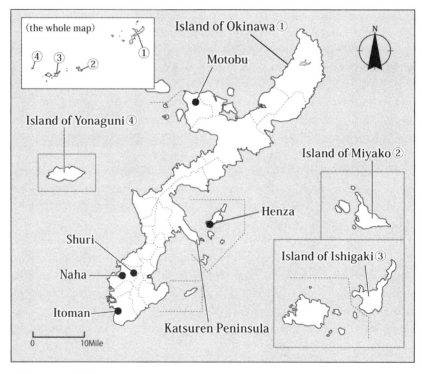

Okinawa, circa 2012.

double-lined fusilier [*gurukun, Pterocaesio diagramma*], scad [*urume, Decapterus muroadsi*], and other kinds of fish into a large tuck net called an *agyā*. Impoverished agricultural areas of the Okinawan chain supplied the ample labor force necessary for carrying out this extremely labor-intensive, although very effective, method.[1] Around the turn of the twentieth century, pole-and-line skipjack tuna fishing was introduced by fishermen from Kagoshima and Miyazaki prefectures and enriched the variation of Itoman fishing. In 1902, more than 60 percent of the full-time fishermen in Okinawa were based in Itoman, which produced more than 60 percent of the profits from fishery products in the prefecture.[2]

The development of Itoman fisheries quickly dried up the fish stocks of the nearby coast and forced the fishermen to constantly travel to find new fishing grounds. After dispersing throughout the prefecture, they moved northward to various shores of mainland Japan along the flow of the Black Current and its branch, the Tsushima Current. As Japan enlarged its sphere of influence by imperial expansion, the fishermen moved southward into Taiwan, Manila, Singapore, Indonesia, Saipan, Borneo, and various shores of the Pacific islands. The warm environments of the South Pacific islands and Southeast Asia, with well-developed coral reefs, resembled their home waters and enabled them to use *agyā* nets and engage in pole-and-line skipjack tuna fishing.[3]

The massive flow of fishermen into foreign waters made Itoman the town in Okinawa that had sent the largest number of people abroad by 1940, but their destinations seemed to be limited to coastal areas of Southeast Asia and the South Pacific. Although a sizable number of residents of Itoman moved to Hawai'i after Okinawans started migrating to the archipelago in 1900, under the leadership of Kyūzō Tōyama, all of them were from farming households and went to work at sugar plantations. Later, some left the cane fields and started fishing.[4] As of 1919, Hawai'i had thirty-three Okinawan fishermen, eighteen of whom operated in Honolulu, twelve on Kaua'i, and three on the island of Hawai'i.[5] Despite their numerical disadvantage as a group within the Wakayama-, Yamaguchi-, and Hiroshima-dominated industry, some of them rose from obscurity and joined the leading figures at sea. Katsuichi Shinzato, for instance, rose to become the owner of several fishing boats. After the war, he strove to revitalize the fisheries of Hawai'i, along with Matsujirō Ōtani, as a vice president of the United Fishing Agency.

Katuichi Shinzato was from Henza, a small island of 610 acres located in the Gulf of Kin, about three miles northeast of the Katsuren Peninsula on the main island. Unlike Itoman, a fishing town, Henza had prospered as a transit point of marine transportation. During its heyday around the mid-1920s, approximately one hundred crafts called *māran-sen* or *yanbaru-sen* carried lumber, firewood, cattle, and horses from the islands of Amami and the northern part of the main island, called the Yanbaru region; on the way back, these boats carried rice, daily necessities, and liquor. While using the sea as a highway connecting the northern and southern parts of the main island, the people of Henza utilized it as a place to catch *okazu* [side dishes]. Unlike the collective fishing style developed in Itoman, the fishing of Henza was done mostly by individuals, mainly on shallow reefs with simple gear. The limited job opportunities and absence of schools beyond the level of compulsory education in Henza urged many young people to leave the island and move to urban areas of Okinawa or mainland Japan for education and work. Significant numbers of them headed for foreign countries, including Taiwan, the Philippines, Singapore, Australia, New Caledonia, the South Pacific islands, the continental United States, Canada, Mexico, and South America. Among these nations and regions, Hawai'i was one of the most popular destinations, and its cane fields absorbed many men and women from the island, including Katsuichi Shinzato. After working in a cane field, he started fishing.[6]

During World War II, Okinawa became a battlefield that saw some of the most bitter fighting in the Pacific theater, claiming approximately 12,000 Americans, 94,000 Japanese military men, including locally recruited Okinawans, and 94,000 Okinawan noncombatants; the total Okinawan casualties of more than 120,000 comprised more than half the total prefectural population. The destructive fires of war, popularly known as the *tetsu-no-bōfū* (iron storm), swept intensely over the once-prosperous fishing town of Itoman and burned to ashes a row of its houses, fishing boats and gear, and considerable other valuable property. Henza evaded a concentrated attack, but repeated air raids on the island sank numerous transportation and fishing vessels.

After the end of the war, Okinawa was separated from mainland Japan and placed under the control of the US military. Soon after the war, Okinawa had been treated as merely a "forgotten rock" of the Pacific. However, the birth of the People's Republic of China on October 1, 1949, and the outbreak of the Korean War in June 1950 boosted the strategic importance of Okinawa as the keystone of the American defense line in the western Pacific. The San Francisco Peace Treaty signed in the middle of the Korean War in 1951 separated Okinawa from Japan for an indefinite period. Major General Josef R. Sheetz, the military governor of the Ryukyus, worked to construct a vast network of permanent military base complexes and facilitated expropriation of the land on a long-term basis.

As a result of the United States Civil Administration of the Ryukyus (USCAR) in 1953, which permitted the seizing of land without the landowners' consent, armed troops forcibly removed occupants from their land and bulldozed their homes and fields, arresting and jailing landowners and other opponents. This oppressive procedure provoked Okinawans and spurred them into *Shima-gurumi tōsō* (the island-wide struggle), a long, bitter struggle against the land acquisition and construction of military bases, crimes committed by US servicemen and perfunctory penalties imposed on offenders, contamination of the soil by the leakage of toxic chemicals, constant noise caused by takeoffs and landings of airplanes at bases built in the middle of residential areas, and other negative consequences of the military presence.[7] Although USCAR promoted democratic administrative procedures by holding elections and nurturing the outward appearance of a civil government, the military often intervened and harassed elected candidates who were critical of American occupation policy and ousted them from office.

In addition to people's anger against USCAR for its dishonest administration, which promoted democratic ideology and practiced military dictatorship, their frustration with the stagnant recovery of the economy and the widening gap of living standards between mainland Japan and Okinawa amplified their criticism of US military control. Gradually, critical voices joined to make strong calls for reversion to Japan.[8] Pacifying the anger of Okinawans became an acute concern of USCAR. It gradually softened its high-handed methods and worked to appease Okinawans by promoting a narrative positioning Americans as "liberators" from Japanese oppression and cultivating an ethnic identity as "Ryukyuan," separate from the Japanese.[9] Fostering affection and respect for Americans seemed essential to redirect the feelings of Okinawans along USCAR's line and continue the presence of US military installations throughout the islands.

Training program for Okinawan fishermen

The US Army noticed the suitability of Okinawans in Hawai'i to be "goodwill intermediaries" between Okinawans and the US military: they shared the same ethnic backgrounds and preserved blood ties.[10] By 1924, when the United States passed the Immigration Act, stopping all further Japanese immigration, the number of Okinawans in Hawai'i had reached 16,534, which was the fourth-largest number following those of Hiroshima, Yamaguchi, and Kumamoto prefectures.[11] Okinawans in Hawai'i did not have an equal relationship with Japanese, however, because Japanese in Hawai'i often discriminated against Okinawans for their dialects and customs and placed them at the lower end of the socioeconomic ladder. The physical and psychological distance between Japanese and Okinawans spread in Hawai'i before the end of the war and caused the latter to welcome the US occupation of the Okinawa archipelago. Without acknowledging the negative influence of US control over the social, economic, political, and cultural affairs of Okinawa and sticking to the despotic image of Japan before democratization, Okinawans in Hawai'i did not believe that the US rule was detrimental, as those in Okinawa claimed.[12]

In the meantime, the hostility between Okinawans and Japanese in Hawai'i gradually declined after the war, indicated by the official joining of the United Okinawan Association of Hawaii (UOA), or Hawai Okinawajin Rengō-kai, and the United Japanese Society of Hawaii, or Hawai Nihonjin Rengō-kai, in 1958; in 1963, the United Japanese Society of Hawaii even elected as its president Shinsuke Nakamine, an issei Okinawan. Despite this increasingly harmonious trend, the US Army attempted to divide them by invoking Japan's past subjugation of Okinawa as its colony. Moreover, the US Army endeavored to bolster an Okinawan identity separate from a Japanese identity by promoting a people-to-people exchange between Okinawa and Hawai'i. With this specific intention, the Ryukyuan-Hawaiian Brotherhood Program started in 1959 under the auspices of USCAR in Okinawa and the US Army Pacific (USARPAC) of Hawai'i. The UOA collaborated with the program. By 1972, when Okinawa reverted to Japan, approximately three hundred Okinawan educators and government leaders from Hawai'i had been to Okinawa, and more than 1,000 government leaders, students, professors, journalists, businessmen, police officers, farmers, and other individuals had come to Hawai'i from Okinawa to receive training in their respective fields. The US Army covered their travel expenses, and the UOA members and many other Okinawans in Hawai'i volunteered to pick up the new arrivals at the airport, hold welcome parties, arrange their training and housing, act as interpreters, and offer recreation programs.[13]

Prior to the inauguration of the Ryukyuan-Hawaiian Brotherhood Program, various relief programs to aid war-torn Okinawa had started immediately after the war. Many large and small relief organizations spontaneously sprang up in the Okinawan community in Hawai'i out of genuine concern for the wretched poverty of their ancestral home. They sent money, clothes, medicine, milking goats, hogs, school supplies, and numerous daily necessities to abate poverty and help reconstruct postwar Okinawa. As part of the relief activities, Chōichi Ige, an Okinawan poultry farmer, and Dr. Y. Baron Goto, then director of the Agricultural Extension Service of the University of Hawai'i, started the Hawaii-Okinawa Farm Youth Training Program in 1952 to strengthen agriculture and alleviate the serious food shortage in Okinawa by transplanting Hawai'i-style agriculture to Okinawan soil. Under the program, trainees worked and lived for six months with host farmers of various ethnic backgrounds, learned the total scope of agriculture in Hawai'i from successful farmers, and exposed themselves to the positive values and practices of the American lifestyle.[14] The US Department of Defense, USCAR, the State Department through its consular office in Okinawa, the Extension Service and the Experiment Station of the University of Hawai'i, the Hawai'i Territorial (after 1959, State) Board of Agriculture and Forestry, and many other agencies were involved in rendering various services.[15] This new endeavor inspired Katsuichi Shinzato, a successful Okinawan fisherman in Honolulu, and his Okinawan friend Sadao Asato, an insurance agent. Shinzato and Asato expanded the program to include fisheries and invited four young fishermen from Okinawa the following year. Since Shinzato was from Henza, three out of four trainees were from Henza, Shinzato's birthplace.[16] In Honolulu, Shinzato took good care of them by feeding them, arranging housing, and making contact with host fishing boats. When the six-month term expired, he provided them with fishing gear as the best possible present for future operations in Okinawa.

The training program for fishermen was, however, not repeated, whereas approximately nine to thirty-one young farmers came every year and received on-the-farm training.[17] One of the major differences between the treatment of farmers and fishermen was the degree of UOA's involvement; the former won wholehearted support from the UOA, whose members offered housing and negotiated with local host farmers,[18] but the latter remained a relatively isolated effort that relied heavily on volunteer contributions by Shinzato and several other local fishermen. When the Ryukyuan-Hawaiian Brotherhood Program began in 1959 as a joint project of the military and the UOA, it firmly integrated the Farm Youth Program into it while excluding an equivalent program for fishermen.

The indifference of the military and the UOA to the reconstruction of the Okinawan fishing industry could not kill Shinzato's vision because Matsujirō Ōtani of the United Fishing Agency and other major figures in the Hawaiian fishing industry became aware of its great potential benefits. Soon, Ōtani and Shinzato started a campaign to restart the fishing training program and reform it by adding a distinctively entrepreneurial view. Unlike the Farm Youth Program, which focused on educational aspects and expected Okinawan farmers to learn, not to earn, in Hawai'i, the proposed project was more business oriented and offered trainees a great chance to make profits. Of course, its most important purpose was to secure a large fishing workforce from Okinawa.

Under the guise of being a training program, the proposed plan was virtually a business contract in which both the United Fishing Agency, the host agent, and participants would mutually benefit while sharing some responsibilities. Unlike the model program for farmers, in which the US Army provided round-trip transportation and local volunteers provided all the necessities of life in Hawai'i,[19] the proposed program required fishermen to pay for airfare, lodging, food, and other expenses. They would borrow money for one-way airfare tickets from the United Fishing Agency and later pay back the debt from their salaries. The term of training was extended from six months to three years, which seemed enough time to pay back all the debts and accumulate profits when they went home. Moreover, a minimum salary of $100 per month was guaranteed regardless of the level of performance.

Ōtani and Shinzato, as well as young bilingual nisei staff members of the United Fishing Agency—in particular Ōtani's second son, Akira, and Frank Goto—played an instrumental role in putting this plan into practice through negotiations with USCAR of Okinawa, USARPAC of Hawai'i, the Hawai'i state and federal government officials, the UOA of Hawai'i, and influential figures in business and political circles in Okinawa. When Masahiro Oyadomari, president of *Ryūkyū Shinpō*, Okinawa's largest newspaper, visited Hawai'i in 1960, Goto made an earnest request to him for help. Moved by Goto's enthusiasm, Oyadomari teamed up with Hikomasa Nagamine, president of the deep-sea tuna fishing association in Okinawa, and appealed to the political and business circles of Okinawa for the realization of the program.[20] In the meantime, Akira Ōtani took full advantage of his personal connections with the US Army, which he had developed during his service as an officer at the GHQ/SCAP, and attempted to mold USCAR's opinion in favor of his plan.[21] This project, thus, symbolized the rise of young nisei replacing their fathers' generation and aggressively advancing new business opportunities with their bilingualism and transpacific networks of acquaintances.

Their years of effort bore fruit. When Katsuichi Shinzato, Frank Goto, Shinsuke Nakamine of the UOA, and an expert on financial affairs of the USARPAC visited Okinawa in 1961, they were finally able to strike a bargain. On September 27 and 28 of that year, Shinzato, Goto, and H. Ted Price from the US Embassy in Tokyo interviewed 130 applicants at the job-placement office in Naha and picked out fifteen for the first group and ten for the second to send to Hawai'i. Shinzato and Goto were careful to select the best fishermen, with advanced skills and good conduct. In particular, they excluded heavy drinkers through careful questioning. Among these interviewees was Tokusaburō Uehara from Itoman, who asserted his abstinence from alcohol and successfully passed the scrutiny.[22]

Tokusaburō Uehara was born in 1936 in Itoman and trained to be a fisherman from a young age. When he saw an advertisement for the program in the newspaper *Ryūkyū Shinpō*, he immediately decided to apply. In particular, the proposed $100 as a guaranteed minimum monthly

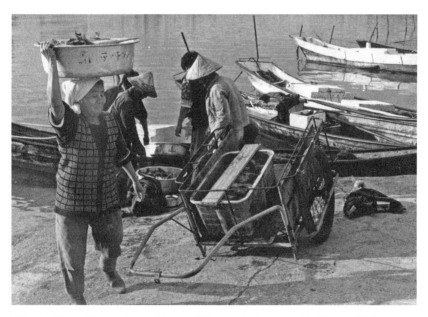

A female fish peddler purchases fish from fishermen in Itoman. Behind her are fishing boats called *sabani*. Date unknown. Photo courtesy of Ken Uehara.

salary was much higher than the average earnings of $30 to $40 per month for high government officials in Okinawa and strongly appealed to him. Hiroshi Nakashima also applied for the program for a similar reason. Born in Taiwan in 1935 and raised on Yonaguni, the southernmost island of the Okinawan chain, Nakashima moved to Itoman in his late teens and started fishing as a profession. When he came home from work and turned on the radio, he heard an announcement for the program, which evoked his childhood dream to go to Hawai'i, a land of fortune. He immediately made up a résumé that misrepresented the facts of his educational background and applied for the program. When he passed the interview, he at last got a chance to make his long-time dream come true.[23]

The fifteen fishermen selected for the first group flew to Honolulu on October 27, 1961.[24] They then stayed on the second floor of the A'ala Market, which Matsujirō Ōtani had remodeled into a residential space equipped with bunk beds and bathrooms, and started working on fishing boats affiliated with the United Fishing Agency. Four men went to skipjack tuna and eleven to tuna longline fishing boats. Their performance during the first three months was more than satisfactory and led to the invitation of the second group, including Tokusaburō Uehara and Hiroshi Nakashima.[25] They left Okinawa on February 16, 1962, with eight other men. Prior to their departure, Akira Ōtani came to Okinawa to pick them up as well as to meet the military governor and enable the continuous operation of the program. On January 19, 1963, the third group, consisting of five men, left for Hawai'i.[26] Out of a total of thirty Okinawan fishermen, more than 70 percent were from Henza, Itoman, and Naha, the capital city of Okinawa.[27] About half of the trainees were married and left wives and children behind at home.[28] In other words, the United Fishing Agency preferentially chose experienced fishermen in their late twenties and thirties so they could make an immediate contribution to the operation rather than inviting young men and training them to be full-fledged fishermen.

In those days, Hawaiian waters also hosted trainees from the trust territories of the Caroline Islands. The Hawaii Tuna Boat Owners Association, affiliated with Hawaiian Tuna Packers, provided them practical instruction in fishing methodology aboard the skipjack fishing boats. These recruits, along with those from Okinawa, were expected to supplement the shrinking population of fishermen, but the local authorities of the

Shintoku Miyagi relaxing on a bunk bed. Miyagi came to Hawai'i in 1962 from Kyan, in the southern part of the main Okinawan island. He worked on a fishing boat affiliated with the United Fishing Agency and stayed in the residential space of the A'ala Market. Photo courtesy of Shintoku Miyagi.

fisheries administration and education gave the programs poor evaluations due to the restrictive three-year limit imposed on the trainee groups. The main ideas prevalent in the Hawaii Fishery Training Committee, an informal gathering started by the Hawaii Area Office of the US Bureau of Commercial Fisheries in 1962, were that training of local recruits would be a far better solution than relying on the temporary and "not always satisfactory" trainees from other areas.[29]

Keenly aware of the urgent need for younger replacements for the retiring older members of the industry, the committee started exploring the possibility of establishing a fishery vocational training program designed to encourage young men to enter fishing as a career. Its plan included comprehensive technical training for high school students who were interested in pursuing careers in navigation, marine diesel mechanics, electronics, and fish catching and marketing. The vision of the committee was not, however, carried out. Of course, Hawaiian Tuna Packers and the Hawaii Tuna Boat Owners Association tried to recruit local youth by making and distributing leaflets among high school students, but few were attracted to fishing. As of 1960, twenty-one skipjack tuna boats were operating in Honolulu, but the number declined to only eleven in 1973. During the same period, local consumers ate the most tuna per person in the United States.[30] The expected influx of labor from outside of Hawai'i was, therefore, critical for the survival of skipjack tuna fisheries and the operation of the cannery.

Without a breakthrough in recruiting local labor, the public and private sectors of the industry inevitably agreed to increase the number of foreign fishermen. Seeing the excellent performance of Okinawan fishermen, the Hawaii Tuna Boat Owners Association turned to Okinawa for its potential labor pool. After the Immigration Act of 1965 abolished the discriminatory national origins quota system, the training program was transformed into immigration in which Okinawans came to Hawai'i with the right of permanent residence.

The involvement of the Hawaiian Tuna Packers in the recruitment of fishermen from Okinawa made it a luxury for the United Fishing Agency to spend time and effort dispatching its staff to Okinawa to carefully screen all applicants and pick only the mature and skilled. Instead, both companies began looking for personal connections to attract as many men as possible. This casual recruitment increased the number of recruits from Itoman and Henza because fishermen from these two places aggressively encouraged their brothers, sons, friends, colleagues, and neighbors to join them. Hisashi Itō came to Hawai'i from Henza in 1969 to replace his *senpai,* a senior fisherman, who said, "I'm coming home, so you should go."[31] The new arrivals included many veteran fishermen as well as amateurs without any fishing experience, such as Kiyoshi Tamashiro, a former taxi driver working at Lucky Taxi and Co. in Itoman. He moved to Hawai'i in 1970 at the recommendation of his supervisor.

> Mr. Ansei, executive director of the company, had a cousin in Hawai'i. The cousin asked Mr. Ansei to find someone to work on skipjack tuna boats. I strongly appealed to him to let me go. I guess about fifteen men went to Hawai'i from Lucky Taxi and Co.[32]

Some of these taxi drivers could not even swim, much less do commercial fishing. The first and foremost motivation for such amateurs to leave home was the excellent financial rewards they expected to earn in Hawai'i. Living in a fishing town, where they had a sense of the occupation, it was relatively easy for them to overcome their initial fears of the dangers of the sea and begin fishing.

In addition to excellent income, adventurous spirits stimulated Okinawan youth to leave home and jump into a new world. Especially in the 1970s, when the serious starvation of postwar Okinawa had already passed into history, some young Okinawans sought a wider world beyond the sphere of their daily routine and saw Hawaiian waters as a place for their personal fulfillment. Tamotsu Ashitomi was one such youth. He was a student in a fishery high school in Naha when he was attracted to Hawai'i. Born into a family of fishermen on Henza in 1951, he grew up with constant encouragement from senior fishermen to go to distant waters. After graduating, he worked as a novice fisherman for one year and headed to Hawai'i in 1971, at age nineteen.[33] Hirofumi Itō, Ashitomi's neighbor and three years younger than he, had also grown up hearing the success stories of his relatives and neighbors already in or returning from Hawai'i and other foreign lands. Although he did not have fishing experience, he went to Hawai'i at age twenty to work on a tuna longline fishing boat.

Prior to their departure for Hawai'i, Tamotsu Ashitomi and Hirofumi Itō had witnessed a tremendous transformation of the landscape and economy of Henza. The development of a road network on the main island of Okinawa had significantly reduced the need for sea transportation, a backbone of the island's economy for centuries. By the end of the 1950s, transportation ships had completely disappeared from the island's waters, a development that embodied the changing position of Henza from the center to the periphery of the Okinawan economy. To reintegrate the island into the land traffic network, the islanders launched the grand enterprise of building an approximately three-mile bridge between Henza and the main island of Okinawa. When they saw the bridge, created by filling the shallows with sandbags and stones, mercilessly washed away by a powerful typhoon and ballooning construction costs far exceeding the estimate, Gulf Oil of the United States approached them. Gulf Oil promised to finish the bridge at its own expense in exchange for installing crude oil terminal stations in Henza. This plan was expected to change the geography of the island tremendously by destroying cane fields and ancestral graveyards as well as damaging local fishing by contaminating the water nearest the beach with debris. After careful consideration by the islanders, Henza accepted the deal.

The advance of a major oil company into Henza around 1970 created many new job opportunities, and the new bridge, completed in 1971, finally enabled people to go to the main island by land.[34] When Hirofumi Itō entered high school, his family and neighbors were able to live a somewhat comfortable, if not luxurious, life. Nevertheless, he decided to leave Henza, "not because I was unable to make a living at home; I just wanted to go out and see the world."[35] He used the opportunity of fishing to satisfy his adventurous spirit.

Okinawan fishermen in Hawai'i

The appearance of Okinawans en masse in Hawaiian waters accelerated dialogue between the sea people of the two cultures, which often resulted in friction and frustration. Upon arriving in Honolulu in February 1962, Tokusaburō Uehara started working on a longline fishing boat affiliated with the United Fishing Agency and soon discovered that Hawai'i lagged far behind in skills and technology:

> In Itoman, fishermen could knit twenty nets to wrap glass floating balls in a day, but, in Hawai'i, fishermen spent two days handling only one ball. Machines on fishing boats in Hawai'i were also out of date and terribly inefficient. Overall, their fishing style was behind by at least twenty years.[36]

As Uehara's surprise indicates, fishermen in Hawai'i in the early 1960s still used the premodern methods brought by issei from Japan, and the overall skills and gear in Hawai'i had not kept up with worldwide advances in fishing technology.[37] In Okinawa, Itoman and other fishing villages had undergone a quick recovery from the devastation of war and adopted the latest techniques and technology of fishing. Solving the serious food shortage and stabilizing Okinawa as quickly as possible was the first and foremost concern for the US military rulers, so they took positive measures to regenerate the fisheries by disposing of military vessels, providing fishing materials and fuel, and supporting ship-building projects. Funds pouring from the US Government and Relief in Occupied Areas program accelerated the mechanization and modernization of boats, diversification of fishing methods, and expansion of infrastructure for cold storage, processing, and distribution. The small *sabani* canoes that once dominated Okinawan waters had been replaced by large vessels equipped with the latest machines and gear by the early 1960s.[38]

Tokusaburō Uehara was the product of this significant postwar development in Okinawan fisheries. A huge gap in the level of skills and methods between himself and his hosts in Hawai'i, mostly nisei Japanese, Portuguese, and Hawaiians, often created discord. Whenever Uehara worked in "his own way," which he believed to be more efficient than following instructions from others, he was scolded, "You are a trainee, and we are your trainers, so follow us." But arguments between Uehara and his "trainers" gradually lessened as the latter started listening to Uehara and adopting his style.[39]

Hiroshi Nakashima (left) and Tokusaburō Uehara (right) came to Hawai'i in 1962 as trainees. Photo courtesy of Tokusaburō Uehara.

In contrast to Uehara, an experienced fisherman, amateurs had to learn everything from scratch, without judging the level of fishing technology in Hawai'i. The strict hierarchy on board, which once allocated to novices the heavy and dirty chores, had mostly loosened by the 1960s, and their only extra duty was to wake up earlier than senior crew members and prepare coffee and breakfast, which did not bother them much. Learning to swim and dive and adjusting to the strong pitching and rolling caused by the large waves in Hawaiian waters were the most formidable challenges they confronted. When Kiyoshi Tamashiro, a former taxi driver from Itoman, started working on a skipjack tuna fishing boat, serious seasickness mercilessly attacked him:

> At first, I had a very hard time handling a long fishing rod. But seasickness annoyed me much more. It took me about three weeks before I got adjusted to the sway of a boat and half a year before I mastered all of the routines.[40]

Unlike Tamashiro, Hirofumi Itō from Henza was not troubled by seasickness. He did not have fishing experience at home, but he had spent much time on board a ferry in his early childhood. His father ran a ferry boat connecting Henza and the main island. In Hawaiʻi, he crewed the *Angel*, a skipjack tuna boat owned by Katsuichi Shinzato; out of its twelve crewmen, a chief engineer, and a captain, eight were from Henza, which made him feel closer to home and reduced the risk of homesickness. But the cold water and wind at sea during the winter pierced the marrow of his bones. Prior to going out to the open water searching for skipjack tuna, his boat went to several locations off Oʻahu, Kāneʻohe Bay, Keʻehi Lagoon, or Pearl Harbor, to catch *nehu*, bait fish. Ito dove into the water wearing only a T-shirt, waited for a school of *nehu* to come to the surface, and surrounded it with a net. Throughout the entire process, cold wind blew on his body in the chilly water.[41] Yukikazu Onaga, a former taxi driver of Lucky Taxi and Co. in Itoman, also had a hard time bait fishing. Whenever he went to Pearl Harbor for bait, the ashes of burned sugarcane, dumped in the bottom of the harbor, caused swelling of his thighs. The huge size of the skipjack tuna, which far surpassed that of the tuna caught in Okinawa, also surprised him. Growing up on the island of Zamami, where skipjack tuna fishing prospered, he had some knowledge and experience of fish and fishing. But the enormous size of the skipjack tuna available in Hawaiʻi far exceeded his expectations and required both tremendous stamina and speed to catch. In his spare time, he practiced lifting and catching a heavy fish again and again.[42]

The hardships of skipjack tuna fishing captured the attention of the National Marine Fisheries Service (NMFS, formerly the Bureau of Commercial Fisheries) in Honolulu, scholars at the University of Hawaiʻi, and other public and private sectors of the fishing industry. Improving the efficiency and reducing the physical burden of fishermen had become one of their research topics. First, the entire process of skipjack tuna fishing took too long. In particular, gathering bait, taking approximately eight hours either at night or during the day, seemed the most likely part to be changed. To reduce the length of the baiting process, sardines and small fish had been introduced into island waters to replace *nehu*, and artificial bait had also been tried. However, these were not as effective as *nehu* and were not adopted by the local fishermen.[43] The NMFS also tried to use purse seines and gill nets, but these attempts ended up failing to catch commercial quantities of fish.[44] The Western Pacific thermocline is deeper than other oceanic areas, enabling the skipjack tuna to swim down and out of the purse seine nets. After all, the procedures for skipjack tuna fishing introduced from Wakayama at the turn of the twentieth century

Fishermen catching bait fish in shallow waters. Photo courtesy of Tamotsu Ashitomi.

turned out to be more effective than any other means. These researchers and scholars could promote only minor technological advances, such as conversion from bamboo to fiberglass poles. Despite being plagued with the time-consuming baiting problems and a labor-intensive fishing style, the landing of skipjack tuna comprised more than 70 percent of the total tonnage, with other fish constituting only 27 percent of the total commercial catch in the state of Hawai'i in 1972.[45] The predominance of skipjack tuna fishing in the industry and better revenues from fresh fish sales silenced complaints among fishermen and even attracted more

Skipjack tuna fishing during the early 1970s. The basic style of skipjack tuna fishing had not changed since before World War II. The only differences were that the fishermen wore helmets, not straw hats, and used fiberglass, not bamboo, fishing poles. Photo courtesy of Tamotsu Ashitomi.

people from other types of fishing. Tamotsu Ashitomi, from Henza, moved to Katsumi Shinzato's skipjack tuna boat *Angel* from a longline fishing boat because he heard that "skipjack tuna fishing is more lucrative."[46]

Hawai'i as a testing ground

Regardless of their fishing techniques and experiences, all Okinawan fishermen confronted a language barrier. In Hawai'i, people spoke Hawai'i Creole English, commonly referred to as "pidgin," which confused them. What's worse, even the local Japanese dialect often confused them because it incorporated Hiroshima and Yamaguchi dialects and accents with which Okinawans were not familiar. Growing up with education in standard Japanese, it took them a while to become accustomed to the local Japanese and technical terms. For Seitoku Kinjō, who was born in Itoman and spent almost all his childhood fishing, the language barriers were extremely hard to overcome:

> I seldom went to school, so I could not speak English in Hawai'i. There was a free English class, but I was busy working and unable to attend it. I still remember how bitterly vexed I was when I had a car accident. I could not understand English, so people laughed at me. I really wanted to run

back to my home. I also had trouble with some Japanese terms for fishing. When someone said *kenken,* I did not understand it, because it was called *hanebiki* in Itoman.[47]

The sense of humiliation and hardship derived from the daunting language barriers were largely offset by the abundance of food and lucrative income, both of which were unobtainable in Okinawa. When Seitoku Kinjō was in Itoman, he always ate *somen* noodles and other humble dishes. In Hawai'i, he could sate himself on beef, pork, and vegetables every day. "I should have come earlier," he thought.[48] The rich diet of Hawai'i deeply impressed young Hirofumi Itō, as well. On his first day in Hawai'i, he was taken directly from the airport to the Flamingo Restaurant in Honolulu for lunch:

> When I saw the combination plate, consisting of three huge fried shrimp and two slices of teriyaki beef steak, I was so surprised that I could only say, "Wow!" Back on Henza, I ate sweet potatoes, somen noodles, and soup every day.[49]

Lavish meals were available onboard as well. When Yukikazu Onaga from Itoman served as a cook as a junior fisherman, his Portuguese captain asked him to prepare a variety of American, Okinawan, and Japanese dishes, including hamburgers, steak, stew, boiled pig's feet, and even sashimi (sliced raw fish). The captain provided him with plentiful provisions, so Onaga and the other crew members were able to eat their fill without worrying.[50]

Together with the abundance of food, the large sum of money paid as a salary also took their breath away. The more experienced and highly skilled fishermen earned from $7,000 to $10,000 each in 1961.[51] Their net earnings could have been much more than the reported figures because they did not disclose their real earnings. Of course, incomes varied by year and season, and differed greatly among captains, owners, and crew members, as well as boats. A more skilled captain and crew earned more than those with lesser skills.[52] In those days, Hawaiian waters were filled with an enormous number of skipjack tuna, which allowed fishermen to chase only schools of huge fish. Hisashi Itō, Hirofumi Itō's cousin, described the abundance of fish "as if we could walk on the school of skipjack tuna."[53] The abundance of fish in Hawai'i brought great benefits to almost all fishermen, including young novices. When Hirofumi Itō

Lunchtime onboard. Photo courtesy of Tamotsu Ashitomi.

received his weekly pay for the first time, he found his pay envelope standing vertically because it contained a thick roll of bills amounting to $500, which was more than five times the monthly payment he received at an automobile company in Yokohama, Japan. The large amount of money made him dizzy with surprise and pleasure.[54]

Itō was not, however, allowed to squander all the money he earned because the United Fishing Agency, his patron institution, managed his income. The company deducted rent, airfare, and other expenses from his earnings and saved the rest, giving him only $20 a month as an allowance; upon request, the company sent money to Okinawa. This system was not endorsed by all the fishermen. "The account books were written in English, which I could not read. I did not know anything about my money in detail. But I could not complain to my patron," said Tokusaburō Uehara. He had no choice but to trust the company and accept this system as the best way to save and send money to his wife and children in Itoman, and not risk wasting his earnings.[55] However, some single young men, including Hirofumi Itō, could not endure an enforced frugal life and often made bargains with Fujiko Nakasone, a nisei woman who was in charge of the accounts at the company:

> When I needed more allowance, I always begged Fujiko-san to withdraw money from my bank account. When I spent a large amount, she scolded me, saying, with a sharp tone, "Save money, and prepare to go back to Henza."[56]

For young fishermen who were inclined to squander their money, Nakasone functioned as a good watchdog to "discipline" them and prevent them from ruining themselves.

Hawaiian Tuna Packers did not take such a paternalistic attitude toward Okinawans working on its affiliated boats. It gave autonomy to the workers by letting them find their own housing and manage their earnings themselves. Some fishermen appreciated the freedom to spend their own money, but others, single young men in particular, found numerous temptations difficult to overcome. Tamotsu Ashitomi saw one Okinawan fisherman become an alcoholic. No one wanted to employ drunkards, and he lost his job and became penniless; eventually, his sister came from Okinawa to Hawai'i to take him home. Although such a case was extreme, single men were inclined to drink all the money they earned.[57] In sharp contrast, Ashitomi did not frequent a bar and worked very hard. His diligence brought him great profits. Within four years, he purchased a condominium in Salt Lake on O'ahu for $51,000. Awhile later, he added a $5,000 car to his possessions. As these stories indicate, Hawai'i was a testing ground for Okinawan fishermen to assess their physical strength and will. Most of them worked hard and accumulated a considerable amount of money, but some could not because they indulged in vices or simply were not good at fishing.[58]

Waiting families in Okinawa

Seitoku Kinjō came to Hawai'i from Itoman in 1973 at age forty-six. Thereafter, he always had his wife and children on his mind. Sending as much money as he could was his utmost concern, so he always declined invitations to go out drinking or smoking. He worked in Hawai'i for twenty-one years. During the first ten years, he did not return home. In the meantime, his wife, Teru, visited him in Hawai'i once, staying for one week. After Seitoku's father and mother passed away, he returned home a couple of times. Overall, Seitoku and Teru rarely saw each other in twenty-one years.[59] What made it possible for them to continue their long-distance marriage for such a long period of time, while hardly seeing each other?

Long-term separations between a husband and wife were fairly common in Itoman, where the chronic absence of fishermen had produced social arrangements and economic practices unparalleled in any other part of the Okinawa archipelago. For instance, a strict separation of budgets between husbands and wives, born of the uncertain and risky environments inherent in the fishing activities of the men, helped women to save up their own money and prepare for the future. Even when a fisherman and his wife lived under one roof, she purchased fish from her husband, went peddling in town, and kept all of the profits for herself, while using her husband's income for family expenditures. Some wives accompanied their husbands to mainland Japan and abroad to peddle their husbands' catch, but only a few wives moved to Hawai'i, which already had a modern fish marketing system. During the years of their husbands' absences, the wives kept busy managing their homes, selling fish, and producing tofu and *kamaboko* (fish pâté). The watchful eyes of neighbors encouraged chastity among the wives.[60]

Teru Kinjō accepted separation from Seitoku as a necessity and led a busy life rearing their five children, taking care of her parents-in-law, and working at a *kamaboko* factory and lunchbox stand. Whenever a spiteful rumor reached her ears that her husband had gotten a woman in Hawai'i, she believed her words to him, "Never do anything that would disgrace our children," would keep him from any vice. To the children, Teru always said, "Your father is working hard in Hawai'i for your sake." Masaru, their eldest son, felt a great sense of loss and responsibility when his father left home; at that time, he was fifteen, old enough to understand the reason Seitoku went to Hawai'i. As Teru expected, Masaru and the other four children had grown up with faith in their father. When Seitoku finally came back to Itoman and joined Teru in 1994, to live in retirement, their reunion after more than two decades of separation made their cohabitation "like the life of a newly married couple."[61] Now Masaru sometimes goes out with Seitoku to shop and do other errands "because I was not able to be dutiful to my father in the past."[62]

The younger generation saw maintaining two households and enduring the loneliness of separate lives as unacceptable. When Tamotsu Ashitomi returned to Henza and married Masayo, his childhood classmate, at age twenty-five, he decided to bring her to his condominium in Salt Lake. His choice was rather uncommon among Okinawan fishermen because the complicated procedures necessary for obtaining permanent residency deterred most of them from bringing their wives and children to Hawai'i. But Tamotsu did not hesitate to make several trips between Hawai'i and Okinawa and go through the complicated formalities. In the meantime, Masayo was not bothered by the idea of going to a new place. She shared her husband's enthusiasm for living abroad, and the presence of relatives and friends in Hawai'i reduced her concerns about life in a foreign land. Tamotsu smiled when thinking about the past: "She married me because she wanted to go to Hawai'i."

Tamotsu and Masayo Ashitomi lived in Hawai'i for three and a half years. While Tamotsu was away fishing, Masayo worked at Tamashiro Fish Market in Pālama, which her relatives ran, and at a bakery, before the birth of their first child. Their life in Hawai'i was comfortable, blessed with nice weather and advanced medical facilities. But the absence of frequent gatherings with neighbors and friends made them feel lonely and somewhat uneasy about their old age. On Henza, all the residents of the island often got together at *eisā* Okinawan dance festivals and athletic meets at schools, which deepened mutual commitment and reciprocity among them. Their aging parents at home also increased their homesickness. In 1985, they finally returned home.[63]

By the time the flow of labor from Okinawa to Hawaiian waters slowed to a stop around the mid-1970s, the total number of Okinawan fishermen had reached about two hundred, including some of those who went back to Okinawa at the expiration of their first three-year labor contract and came back to Hawai'i later with immigrant status. They reduced the shortage of local fishermen and offered a solution, if temporary, for the fisheries' most acute problem. Their numbers shrank when the majority of fishermen returned home within three to five years to reunite with their wives, children, and parents. Some of them continued fishing in Okinawa, utilizing skills they had obtained in Hawai'i, while others returned to their old land jobs. During their stay in Hawai'i, they did not develop deep ties to local Okinawan society. Fishermen generally spent most of each day onboard vessels, and the rest at dormitories or houses shared by five or six men. Their work-oriented lifestyles kept them together a great deal and lessened their chances to become attached to the local communities in Hawai'i. Most single men who married local women and settled in their adopted land left the fishing boats and got jobs on shore, such as running fish markets and sushi restaurants. Afraid the separation from the rest of the society would reduce his chances of getting married, Hirofumi

Itō left the skipjack tuna boat on which he had spent about a decade. After that, he married a local Japanese woman and supported his family as a tour bus driver. "Fishing has become nothing more than a memory of my youth," Itō said.[64]

While the number of Okinawan fishermen was dwindling, fishermen of other ethnic groups gradually entered Hawaiian waters. Soon, they would open a new era for the seascapes of Hawai'i, and simultaneously end the domination of Japanese fishermen, which had lasted for more than seven decades.

Epilogue

Fishing operations after Japanese domination

> It is very difficult to make a living by fishing because fuel oil is expensive and the closed season for certain kinds of fishing is very long. I'm already seventy years old and semiretired, but my ship is still in good condition and I want to go fishing as long as possible. Sometimes I miss Okinawa, but my cousin, a fisherman in Okinawa, says that it is hard to manage there because of the rise in oil prices. So, I will keep operating here for at least ten more years. I don't want to fall behind in the industry, and I always want to pioneer new fishing methods and explore new fishing grounds.[1]

On the deck of his boat, *Lisa I,* at anchor in 'Ilikai Yacht Harbor in Honolulu, Hiroshi Nakashima honestly expressed his feelings. Since he came to Hawai'i in February 1962 as a member of the second group of Okinawan fishing trainees, at age twenty-seven, he has spent most of his days in Hawaiian waters. Unlike many of his former Okinawan fishermen colleagues, who returned home or took up land jobs, Nakashima remained onboard and kept fishing, while observing significant changes in the structure and demography of Hawai'i's commercial fishing since the 1970s.

One of the most noticeable changes in the industry was the rise of the island of O'ahu as the primary location of fish-processing facilities in Hawai'i. By the late 1970s, O'ahu, especially Honolulu, had grown to include more than 80 percent of the state population. The sizable presence of consumers attracted a much larger amount of marine products than did any of the other islands in the archipelago. In 1978, O'ahu received 84 percent of the state's total fish catch, while the island of Hawai'i accounted for 8 percent, and the combined haul of Maui, Lānaʻi, Molokaʻi, and Kauaʻi constituted the remaining 8 percent of the total catch. Among more than twenty-one fish landing points on Oʻahu, Kewalo Basin in Honolulu accounted for approximately 95 percent of all the state's landings due to the availability of infrastructure and facilities for distribution and processing.[2] In 1979, the United Fishing Agency and its Honolulu Fish Auction moved to Kakaʻako from Aʻala and became a new landmark in this fishing-centered community.

The other striking characteristic of Hawaiian fisheries during the late 1970s and 1980s was the diverse ethnic background of the commercial fishing population. Fishermen of various ethnic groups, in particular, Korean, Vietnamese, and white, gradually took the place of Japanese and Okinawan fishermen. Later, Tongan, Marshall Islander, Filipino, and Indonesian fishermen joined the local fleets, while fisherwomen, albeit in much smaller numbers than men, advanced to the sea and contributed to the diversity of the commercial fishing population in Hawaiʻi.[3] In Hilo, on the island of Hawaiʻi, which once prospered as a home port for hundreds of Japanese fishing boats, serious aging and shrinking of the population of Japanese fishermen posed a significant problem for the industry. Unlike Honolulu, Hilo did not introduce Okinawan labor. Worse, a devastating tsunami in 1960 swept away the coastal district of Hilo Bay and utterly demolished fishing boats, which accelerated the dismantling of Japanese fishing fleets. In the meantime, Filipino, Hawaiian, and white fishermen increased their presence, and grew to be the principal suppliers of fresh fish in Hilo. "From our [Japanese] point of view, their fishing skills are not good. But securing a haul is much more important, and we have to appreciate them for supplying fresh fish," said Zenzō Kanai, an executive of Suisan Co. in Hilo.[4]

The extensive integration of new fishermen and fisherwomen into the seascapes of Hawaiʻi has reinvigorated deep-sea longline fishing (tuna, billfish, and swordfish), which had flourished soon after the war but had declined by the late 1970s. By aggressively introducing modern, long-range vessels from the continental United States, the newcomers have expanded the longline fishing fleets into the dominant group in the islands' fisheries. Moreover, the discovery of the Northwestern Hawaiian Islands lobster fishery after 1980 and the rediscovery and development of its bottom fish (snapper, grouper, and jack) were precious additions to the local industry. The troll fishery targeting blue marlin, yellowfin tuna, mahimahi (dolphinfish, *Coryphaena hippurus*), ono (wahoo, *Acanthocybium solandri*), and skipjack tuna became popular among both full- and part-time commercial fishermen.[5]

In contrast to these emerging and reemerging fishing methods, mostly promoted by new faces, the sampan pole-and-line fisheries, which were introduced by the Japanese in the early twentieth century and remained the largest commercial fishery in Hawai'i, contributing from 40 to 60 percent of the annual commercial landings throughout the 1960s and 1970s, dwindled markedly during the 1980s.[6] The primary stimulus for the decline was the closing of Hawaiian Tuna Packers in Kaka'ako. During its peak period in the early 1950s, Hawaiian Tuna Packers had more than thirty-five affiliated boats and nearly four hundred employees supplying raw materials for canning.[7] It took pride in its Coral brand as the only canned tuna in the world consistently packed from fresh tuna, but its quality-oriented production inevitably caused various problems deriving from an unstable supply of local fresh fish and bitter competition with low-cost rival products. On top of economic restructuring in the global tuna industry, numerous health measures imposed by the Federal Environmental Protection Agency drove its management further into a corner.[8] In 1984, Hawaiian Tuna Packers finally ended its long history, after starting as a joint business between the Japanese and Americans in 1922 and functioning as the backbone of the local fisheries until its demise.

The closure of the cannery drastically reduced the local demand for skipjack tuna, decimated the sampan fleets, and spurred the remaining Okinawan crew members homeward. By the mid-1980s, the number of boats in the fishery had fallen to nine (from fifteen in 1971); in 1991, six boats were active, four on a full-time basis.[9] By September 2008, the number of Okinawan fishermen working full-time on O'ahu was less than ten, and only two sampan boats operated, the *Kula Kai* and the *Nisei*, based at Kewalo Basin. In 2009, Captain Fukuo Kinjō, from Itoman in Okinawa, retired, and his *Kula Kai*, built in 1947 by Seiichi Funai from Wakayama, ceased to operate. Masami Shinzato, an Okinawan crew member of the *Nisei*, now the only remaining sampan on O'ahu, explained the reason he chose to stay in Hawaiian waters:

> I came here in 1969 from Henza at the invitation of my uncle, Katsuichi Shinzato, a leading figure of the local Okinawan fishermen. After engaging in tuna longline fisheries for a while, I moved to a more lucrative skipjack tuna pole-and-line fishing boat. Originally, I did not intend to permanently settle in Hawai'i, but the dropping exchange rate of US dollars to

the Japanese yen after the reversion of Okinawa to Japanese control in 1972 diminished my assets and shattered my dream to go home as a successful man.[10]

In 2010, the *Nisei* had four Okinawan crew members, including Shinzato, and navigated under the command of an Okinawan captain from Itoman, Hiroshige Uehara. Shinzato said, "All of us, except one, married local women and started families here. Thanks to the warm temperatures and absence of typhoons, it is relatively easy to work in Hawai'i. Even if we return to Okinawa, we have no place to live there anymore."[11]

An Okinawan fisherman's family and the contemporary fishing industry in Hawai'i

Shinzato's phrase, "I have no place to live there anymore," was repeated by Takeko Nakashima, the wife of Hiroshi Nakashima. Born in 1939 on Tarama, a small island of approximately 7.6 square miles in the southern part of the Okinawa archipelago, Takeko married Hiroshi and started a family in Naha, the prefectural capital. Subsequent to Hiroshi's departure for Hawai'i in 1962, they lost their house in Naha due to troubles with their relatives. A large amount of money sent from Hiroshi relieved Takeko and their three children to some degree, but a sense of despair over their relatives' betrayal and the prospect of a new life in a new place encouraged her to move to Hawai'i with the children and join Hiroshi.

In Hawai'i, another trial awaited her. As soon as she was settled in Honolulu in 1974, Takeko started working at a fish retail shop run by an Okinawan couple, and learned how to weigh, buy, and sell fish. After only one week's apprenticeship there, she stood alone at the counter of her own shop, in a leased space in the O'ahu Market. Back in Naha, she had worked in various places, but she never had a job related to the fishing business. Moreover, a language barrier and numerous physical and mental stresses of living and working in a new environment amplified her distress:

> Soon after I started the fish shop, I suffered from an unknown illness and my fingers were paralyzed. My doctor diagnosed it as an incurable, unknown disease. So I took a break for a couple of weeks in Alaska, leaving my husband and children behind. Thanks to a relaxed vacation in Alaska, my paralysis completely disappeared, and I recovered my health. Since that incident, I have been healthy and have never seen a doctor. But the pressure of speaking English has always annoyed me. When I was growing up on

> the island of Tarama, I was absent from school very often. A serious food shortage forced us children to work along with our parents. I did not study English well. Through working here, I've picked up several English words relevant to my business, but, even now, I don't understand other English words.[12]

In spite of her poor English, Takeko gradually got the knack of her fish retail business. In 1985, she purchased the largest space at the Oʻahu Market. In the following year, Hiroshi built his own boat, *Lisa I*. His excellent fishing skills and diligence won the trust of a nisei banker who financed him. His relatives in Okinawa, including a ship carpenter, came to Hawaiʻi and helped build *Lisa I*. At that time, Hiroshi and Takeko bought a house in Kalihi, on the outskirts of Honolulu. The large expenditure within a short period of time pressed them to work harder than before, without allowing them to make even a short visit to Okinawa. Takeko kept her shop open 365 days a year, and Hiroshi went out to sea on an almost nonstop basis.

> After I built my boat, I was completely absorbed in fishing and desperate to pay back the huge debts. Fortunately, fish were abundant, and I didn't mind working twenty hours a day. The more I worked, the more I earned. Within seven years, I cleared all of my debt.[13]

Because of the constant absence of their parents, the Nakashimas' three children "grew up by themselves."[14] Lisa, their only daughter, looked back at her childhood: "I rarely saw my father. I really missed my dad."[15] Because both of her parents were always away from home, she spent most of her time with her two brothers or other children:

> I was a latchkey kid, hanging a key from my neck and getting into my house with it. But there were many such children in my neighborhood at that time, and I did not feel lonely or neglected. I didn't think my parents were unusual.[16]

Her great sympathy and respect for her hardworking parents prevented Lisa and her brothers from straying from the right path. Immediately after graduating from high school, Lisa started helping her father as a fisherwoman for about two months. Her brother Allen also worked on the *Lisa I*, but neither of them persevered in the fishing business. Nevertheless, Lisa and her brothers still indirectly support their father by helping at Takeko's shop, which sells his catch. "As long as my father engages in fishing, my mother will keep running our shop. He deeply loves the sea, and she loves people," says Lisa.[17]

The business style of the Nakashima family resembles the economic practices of Itoman and many other fishing communities in Japan, where men went out to sea to fish and their wives sold their catch. Even today, some places still preserve this custom. The trajectories of their individual experiences in Hawai'i from the mid-1970s until today mirror the history of Japanese fisheries since the turn of the twentieth century. Hiroshi's constant efforts to understand the local marine fauna and search for new fishing styles remind us of his predecessors from Wakayama, Yamaguchi, Hiroshima, Okinawa, and other parts of Japan. Behind the rise of the Japanese fishermen were the significant contributions of their wives, who were heavily involved in local fish distribution channels and processing, both of which were also indispensable for the expansion of the fisheries.

Even after the end of Japanese and Okinawan domination at sea, their presence remained significant in the processing and distribution sectors of the industry in the following decades. James N. Okuhara, the founder of Okuhara Foods, Inc., is one such case. Okuhara, a nisei Okinawan, opened a small delicatessen at the A'ala Market in 1947. Later, he concentrated on *kamaboko* production. After Matsujirō Ōtani's *kamaboko* plant closed down during the war due to a shortage of raw materials, small *kamaboko* factories sprang up one after another in postwar Hawai'i and competed with each other. Okuhara did not have previous experience in treating fish, but he overcame many ups and downs in the business and expanded his products to become the leading *kamaboko* brand in the state. Okuhara, now eighty-five years old, still goes to work, sitting at his desk in the company's office next to his nisei Okinawan wife, Sueko. They say, "It is better for our health to come here and work than to stay at home and watch TV."[18]

During the 1970s, in fish markets in Hawai'i, where fish for sale were displayed on the counters of rented or leased stalls, approximately 71 percent of all fish dealers were of Japanese and Okinawan ancestry; 14 percent, Chinese; 11 percent, Hawaiian; and 4 percent, white. The rise of supermarkets since the late 1940s has tremendously affected the local fish distribution system and deprived small retailers of their shoppers. However, old-style mom-and-pop shops, including Takeko Nakashima's, continue to exist, patronized by loyal customers. Peddling, the oldest style of fish sales in Hawai'i, also remains a part of the local landscape, although the number

Takeko Nakashima in front of her fish shop with her daughter, Lisa, and granddaughter, Laurally. Photo by the author.

of licensed fish peddlers shrank from seventy-five during the presupermarket era to only fifteen by the late 1970s. At the turn of the twentieth century, fish peddlers put fish in baskets and walked around plantations; about one hundred years later, they travel through urban and rural areas in automobiles, modified vans, or pickup trucks, and sell fish and other food items, such as red meat and vegetables.[19]

The fish auction, introduced to Hawai'i by the Japanese during the 1910s, is still thriving today as a core intersection of sea and society, offering the principal outlet for fresh fish. The Honolulu Fish Auction of the United Fishing Agency, in particular, has been an anchor of the fishing community on O'ahu. Takeko Nakashima said:

> Until about ten years ago, I went to the Honolulu Fish Auction early every morning. Other Japanese fish dealers gave me a ride because I didn't drive. At bidding, we were rivals, exchanging prices like a quarrel. After the auction, we came back together in good harmony. It was really fun. In those days, I was the youngest retailer around here. Now we are all in our seventies and eighties.[20]

Now Takeko no longer goes to the auction. Instead, she places orders with wholesalers over the phone. But she knows that the auction still serves as a special site for setting prices, exchanging information, and socialization among fish dealers and retailers, as it did almost a century ago. The situation was the same in Hilo, where the Suisan Co. ran a fish auction at the mouth of the Wailoa River. It attracted throngs of local fish retailers, dealers, and many tourists every morning except Sundays. Despite the repeated tsunami tidal waves, it came back to life again and again. Strict public regulation of fishing operations and handling of the day's catch, together with indictments for violations of the Hazard Analysis and Critical Control Point program of the US Food and Drug Administration, exhausted its management and finally led to the termination of all auction activities in 2001.[21]

Whereas Hilo lost part of its great heritage from the Japanese fishing pioneers, the district of Kaka'ako in Honolulu went through dramatic changes at the turn of the twenty-first century. After the shutdown of Hawaiian Tuna Packers, the Hawai'i Community Development Authority promoted the redevelopment of Kaka'ako. Residential areas once inhabited by many Japanese fishermen and their families were bulldozed and replaced by buildings for commercial and recreational uses, including the University of Hawai'i's John A. Burns School of Medicine. In 2004, the Honolulu Fish Auction and the headquarters of the United Fishing Agency moved to the state's new commercial fishing village at Pier 38 in Honolulu Harbor. This village was planned to bring together fishing boats, a fish auction site, fishing supply and support companies, seafood wholesalers, and restaurants at one location. Former governor Ben Cayetano, a champion of this project, envisioned the future of the fishing village: "Commercial fishing companies would be centered with an eye to selling fish products and also creating an attraction for local people and tourists—much like the world-famous Tokyo Fish Market in Japan."[22]

As Cayetano described, this place was modeled on the Tsukiji Fish Market, a major hub for the Japanese fishing industry, where numerous professionals and amateurs flock together to sell, purchase, and consume marine products brought from all corners of the world. The scale of the new fishing village in Honolulu is far smaller than that of Tsukiji, but it has steadily grown to be a center of the state fishing industry. In 2006, Honolulu received approximately 72 percent of the total marine catch in Hawai'i. The majority of it was sold through the Honolulu Fish Auction

The Honolulu Fish Auction of the United Fishing Agency. Photo by the author.

and catered to local customers, while some was flown to other islands of Hawai'i and the mainland United States, and exported abroad.[23]

Fish consumption continues to be an indispensable part of everyday life in Hawai'i, and a key ingredient of regional cuisine is, of course, locally produced seafood. Some of the culinary culture dates back to traditional Hawaiian use of near-shore and reef species, and other parts come from the Japanese tradition of eating tuna and deep-water fish. In the past, the local fish market was classified by ethnic group preference: Chinese liked white flesh appropriate for steaming or cooking Chinese-style; Japanese were particularly fond of red-fleshed fish suitable for eating raw as sashimi; Samoans were accustomed to the reef and inshore fish; Filipinos bought reef fish that were eaten in the Philippines; Hawaiians probably ate a greater variety of fish than did members of any other ethnic group; whites generally ate very little fresh fish caught locally, and liked frozen or chilled filleted fish.[24]

In the 1970s, the ethnic composition of the state seafood-eating population began to significantly change, mainly due to the influx of heterogeneous immigrant populations and the expanding tourist industry. The consumption of white-fleshed fish, in particular mahimahi, rose among white tourists from the mainland. The shift toward greater seafood consumption among the health-conscious white and resident population of Hawai'i contributed to greater awareness of and preference for fresh seafood in general. As a result, catches and prices of mahimahi and ono, the local production of which was low in the 1960s, began to rise dramatically in the 1970s.[25] Simultaneously, the Japanese style of eating fish raw spread beyond the ethnic cluster. Nobuo Tsuchiya, a former Japanese sushi chef in Honolulu, explained how sushi and raw fish were accepted by the white population:

> After graduating from high school, I started my career as a sushi chef in Tokyo. I came to Hawai'i in 1971, at the invitation of a sushi restaurant in Honolulu when I was twenty-two years old. In those days, most of our customers were issei, nisei, and Japanese residents working in Hawai'i. About twenty years ago, I started serving special sushi rolls arranged for white people, because they started eating sushi. They usually came to eat sushi for the first time accompanied by their Japanese friends. At first, they ate boiled shrimp sushi, and, later, they moved to try raw fish sushi and became familiar with its tastes.[26]

After quitting the restaurant, he started a wholesale company specializing in tuna and saw a rapid increase in the amount of domestic as well as imported fresh and frozen fish bringing greater variety to regional seafood cuisines. Nowadays, a person in Hawai'i eats forty-two pounds of fish per year, which is nearly three times more than the national average.[27]

A new chapter in Hawai'i-Japan fishing history

Today, the great technological developments of fishing vessels and improved freezing systems have made it possible for fishermen from Japan to travel directly to and operate in the central Pacific and return home carrying a full load of fish. Even if they have a chance to visit Hawai'i en masse, they no longer form a community and do business as their predecessors did a century ago. They stay only briefly and develop little association with the local fishing community. Hisao and Shizue Shimizu sketched the attenuated interaction between the two groups:

> During the 1950s, there were not many issei Japanese fishermen remaining in Hawai'i; many of them returned home, or passed away. But we were busy treating many guests from Wakayama. They came here on university training ships, Hayashikane's *Banshū-maru,* and many other ships. The Wakayama Kenjinkai [prefectural society] and Ōtani's company often asked us to entertain Japanese crews, especially those from Wakayama. Even if we didn't know them at all, they still counted on us because we are from Wakayama. We took them on an island tour and held a welcome party. As a token of gratitude for such entertainment, we received a whole *shibi* [tuna], which we could not finish because it was so big. Today, no one visits us. No relatives even come to see us anymore.[28]

Hisao engaged in fishing with Matsutarō, his father-in-law, on the *Kinan-maru,* after coming to Honolulu from Wakayama in 1955. When Matsutarō passed away about two decades later, Hisao sold the *Kinan-maru* and took a job in termite control and, later, carpentry. The hardships inherent in fishing operations, in particular, chronic seasickness and long-term separation from his wife and children, drove him to give up fishing. Awhile later, he heard that the *Kinan-maru* had run aground and sunk near the island of Moloka'i.

The loss of the *Kinan-maru,* a token of Matsutarō Shimizu's recovery from sizeable wartime losses and hardships, seemed to symbolize the vicissitudes of Japanese presence in Hawaiian waters. The Japanese came to Hawai'i around the beginning of the twentieth century, prospered as the leaders of the fishing industry in the following decades, suffered a crushing blow during World War II, regenerated and revitalized themselves with new blood from Okinawa after the war, and gradually diminished with the rise of new fishing populations. Due to the elusive and migratory nature of fish, fishermen have a strong propensity to constantly move to better fishing grounds. If one place ceases to be attractive, they unhesitatingly leave and head to another one. Hawaiian waters used to be rich in marine resources, but overfishing has depleted fish stocks and diminished their lucrativeness.[29] In that sense, it is nothing but a natural consequence that the Japanese fishing fleets, which prospered in Hawai'i at a certain point in time, faded away at another moment in history, moving like the ebbs and flows of the tide.

Even after Japanese fishermen retreated from Hawaiian shores, the historical narrative of Hawai'i-Japan relations in the fisheries continues, adding new chapters every day and every year. The great sympathy that people in Hawai'i expressed toward the victims of the *Ehime Maru* accident is an indispensable part of this story. In February 2001, the Uwajima

Ehime Maru memorial at Kaka'ako Waterfront Park. Photo by the author.

Fisheries High School training ship *Ehime Maru,* from Ehime Prefecture in Japan, was hit and sunk by the USS *Greeneville,* resulting in the loss of four students and five crew members. When the *Ehime Maru* Memorial was established at Kaka'ako Waterfront Park, the St. Louis School's Japanese Club were the first to volunteers to clean it, and have been doing so every third Saturday of the month since the unveiling ceremony. After the tragedy, the people of Hawai'i and Ehime have held and attended memorial services at every anniversary, and jointly initiated and perpetuate goodwill between them by exchanging youth baseball teams, starting internship programs for University of Hawai'i students to teach English and learn business in Ehime, and kicking off sister school programs between Hawai'i's Kawananakoa Middle School and Ehime Uwajima Minami Junior High School. These programs continue to this day.[30] Through commemorating the victims in a constructive way, the people of Hawai'i and Ehime are making lofty efforts to remind others of the tragedy and strive for maritime safety, which is a permanent desire for all people whose lives are closely associated with the sea. The dialogue continues between Hawai'i and Japan over the sea, the fruitful cradle of fishing culture, and a distinctive space for exchanges of peoples, ideas, equipment, commodities, and other precious items.

Notes

Introduction

Epigraph: Ishigaki Ayako, *Ai to wakare* [Love and farewell] (Tokyo: Kōbunsha, 1958), 26. All translations from Japanese sources are my own.

1. Theodore C. Bestor, *Tsukiji: The Fish Market at the Center of the World* (Berkeley: University of California Press, 2004), 29.

2. Mark R. Peattie, *Nan'yō: The Rise and Fall of the Japanese in Micronesia, 1885–1945* (Honolulu: University of Hawai'i Press, 1988), 1–2.

3. *Gishi-wajinden, Gokansho-tōiden, Sōsho-wakokuden, Suisho-wakokuden* (Tokyo: Iwanami Shoten, 1951), 43.

4. Katrina Gulliver, "Finding the Pacific World," *Journal of World History* 22, no. 1 (2011): 90.

5. Franklin Odo, *No Sword to Bury: Japanese Americans in Hawai'i during World War II* (Philadelphia: Temple University Press, 2004), 6.

6. Such studies include Eiichiro Azuma, *Between Two Empires: Race, History, and Transnationalism in Japanese America* (Oxford: Oxford University Press, 2005); Yuko Konno, "Trans-Pacific Localism: Prewar Village Ties That Connected Taiji, Wakayama, to Terminal Island, California," *Journal of American and Canadian Studies* 29 (2011): 29–57.

7. Yoneyama Hiroshi, "Kan-Taiheiyō chiiki ni okeru Nihonjin no idousei o saihakken suru" [Rediscovering the transpacific mobility of the Japanese], in *Nikkeijin no keiken to kokusai idō: Zaigai Nihonjin, imin no kingendaishi* [Transnational Japanese mobility in the modern era: The experiences of overseas Japanese and their descendants], ed. Yoneyama Hiroshi and Kawahara Norifumi (Tokyo: Jinbun Shoin, 2007), 9–23.

8. Gary Y. Okihiro, *Cane Fires: The Anti-Japanese Movement in Hawai'i, 1865–1945* (Philadelphia: Temple University Press, 1991), 18. Dennis M. Ogawa's pioneering book of Japanese immigrant history in Hawai'i, for example, focuses on the stories of plantation communities without paying attention to the seaside communities. Shiramizu Shigehiko's work on the multicultural environments of Hawai'i affirms that the "loco" or local residents of Hawai'i share the same cultural roots in the sugarcane

plantations. This land-centric trend is evident in a study that traces the origin of "Asian settler colonialism" to the colonial apparatus of the sugar plantations. See Dennis M. Ogawa, *Kodomo no Tame ni: For the Sake of the Children* (Honolulu: University of Hawai'i Press, 1978); Shiramizu Shigehiko, "Taiheiyō no rakuen saikō" [Reexamining the paradise of the Pacific], in *Tabunka shakai Hawai no riaritī: minzoku-kan kōshō to bunka sōsei* [The vital realities of multicultural Hawai'i: Contention, conciliation, construction], ed. Shiramizu Shigehiko (Tokyo: Ochanomizu Shobō, 2011), 8–10; Candace Fujikane, "Introduction," in *Asian Settler Colonialism: From Local Governance to the Habits of Everyday Life in Hawai'i*, ed. Candace Fujikane and Jonathan Y. Okamura (Honolulu: University of Hawai'i Press, 2008), 17–21.

9. Ishikawa Tomonori, *Nihon imin no chirigakuteki kenkyū* [The geographical study of Japanese immigrants] (Ginowan: Yōju Shorin, 1997), 465.

10. Okihiro, *Cane Fires,* 186.

11. Japanese historian Ikumi Yanagisawa's study of a picture bride has refuted a popular discourse in which she was supposedly enslaved by a patriarchal household system at home and forcibly shipped off to a strange land to marry a strange man. Through elaborate analysis of interviews with picture brides, Yanagisawa revealed that many of them made the final decisions themselves to come to the United States and marry men with whom they had nurtured relationships of trust and feelings of intimacy through correspondence prior to marriage. For some picture brides, marriage was an excuse to come to the United States and explore a new, wider world. Yanagisawa Ikumi, "'Shashin hanayome' wa 'otto no dorei' dattanoka: 'shasin hanayome' tachi no katari o chūshin ni" [Was a 'picture bride' a slave of her husband? Narratives of picture brides], in *Shashin hanayome, Sensō hanayome no tadotta michi: Josei-iminshi no hakkutsu* [Crossing the ocean: A new look at the history of Japanese picture brides and war brides], ed. Shimada Noriko (Tokyo: Akashi Shoten, 2009), 47–85.

12. Gary Y. Okihiro, *Margins and Mainstreams: Asians in American History and Culture* (Seattle: University of Washington Press, 1994), 84.

13. These works include Arrell Morgan Gibson with John S. Whitehead, *Yankees in Paradise: The Pacific Basin Frontier* (Albuquerque: University of New Mexico Press, 1993); Mansel G. Blackford, *Pathways to the Present: U.S. Development and Its Consequences in the Pacific* (Honolulu: University of Hawai'i Press, 2007); Andrew F. Smith, *American Tuna: The Rise and Fall of an Improbable Food* (Berkeley: University of California Press, 2012); Mansel G. Blackford, *Making Seafood Sustainable: American Experiences in Global Perspective* (Philadelphia: University of Pennsylvania Press, 2012); Ryan Tucker Jones, "Running into Whales: The History of the North Pacific from below the Waves," *American Historical Review* 118 (2013): 349–377.

Chapter 1: Passage to Hawai'i

1. Kawaoka Takeharu, *Umi-no-tami: Rekishi to minzoku* [People of the sea: Their history and customs] (Tokyo: Heibonsha, 1987), 109.

2. Sea tangles, or *konbu* in Japanese, is edible kelp, widely used in Japan to make soup stock.

3. Tajima Yoshiya, "Kita no umi ni mukatta Kishū shōnin" [The Kishū merchants who went to the northern lands], in *Nihonkai to hokkoku bunka* [The Sea of Japan and northern culture], ed. Amino Yoshihiko (Tokyo: Shogakkan, 1990), 374–426.

4. Morimoto Takashi, *Tōwachō-shi: Kakuron-hen dai sankan gyogyōshi* [History of Tōwachō: vol. 3, The record of fisheries] (Suō-Ōshimachō: Tōwachō Yakuba, 1986), 36–37.

5. Yoshida Keiichi, *Chōsen suisan kaihatsu-shi* [On the development of the Korean fishery] (Shimonoseki: Chōsuikai, 1954), 88.

6. Miyamoto Tsuneichi, *Setonaikai no kenkyū* [The study of Seto Inland Sea] (Tokyo: Miraisha, 1965), 359.

7. Ibid., 358.

8. Yoshida, *Chōsen suisan kaihatsu-shi*, 106–107.

9. Miyamoto Tsuneichi, *Tsushima gyogyō-shi* [The history of fisheries in Tsushima] (Tokyo: Miraisha, 1983), 230–305.

10. I use "coastal communities," "coastal villages," and "fishing villages" interchangeably in this book, although the concepts are slightly different. Historian Arne Kalland argues that, during the Tokugawa period, "coastal village" denoted a village with special relations to the feudal authorities that was administered by a separate department; "fishing village" denoted a village with access to the sea, where some of the inhabitants worked as fishermen on fishing boats registered in the village. Arne Kalland, *Fishing Villages in Tokugawa Japan* (Honolulu: University of Hawai'i Press, 1995), 23.

11. Miwa Chitoshi, "Hiroshimaken-ka Setonaikai mikaihou buraku gyofu no senkai shutsuryō" [Going out fishing in Korean waters of *buraku* fishermen in Hiroshima prefecture of Setonaikai], *Gyogyō Keizai Kenkyū* 21, no. 2 (1975): 43–51.

12. Morimoto, *Tōwachō-shi*, 131–183.

13. Ibid., 181.

14. Nagasaki Prefecture sent 58 boats, Oita 45, Kagawa 40, Okayama 38, Ehime 21, Kagoshima 14, Fukuoka 11, and Kumamoto 10. Kim Byungchul, *Ebune no minzokushi: Gendai Nihon ni ikiru umi no tami* [The history of sea nomads: The sea people living in contemporary Japan] (Tokyo: Tokyo Daigaku Shuppan, 2003), 88.

15. Takeda Naoko, *Manira e watatta setouchi gyomin* [Fishermen of Seto Inland Sea who went to Manila] (Tokyo: Ochanomizu Shobō, 2002), 159.

16. Kim, *Ebune no minzokushi*, 89. These Japanese engaged in various types of fishing by utilizing a variety of gear and methods. Those from Nagasaki Prefecture, for instance, used special anglerfish nets to take advantage of the tide, as they did in the Ariakekai, which had the most extreme tides in Japan. Although diving was prevalent off the coast of Cheju and adjacent areas of the south coast of Korea, the fishermen from Nagasaki, Tokushima, Ehime, Oita, Hiroshima, Yamaguchi, Hyogo, Okayama, and Wakayama prefectures distinguished themselves by engaging in diving-apparatus fisheries, catching sea cucumber, abalone, and other sea creatures. An abundance of whales in Korean waters attracted Japanese fishing companies from Kagawa, Nagasaki, Yamaguchi, and Mie prefectures that conducted Norwegian-style whaling. It should also be noted that the development of sardine fisheries accompanied the increase in manufacturing dried sardines and sardine-oil fertilizer, products that were sent to Japan. By introducing many fishing techniques and implements they had developed in their respective prefectures, Japanese fishermen expanded modern commercial fisheries in Korea.

17. Morimoto, *Tōwachō-shi*, 196–213.

18. Yokemoto Masafumi, "Senzen-ki Taiwan ni okeru Nihonjin gyogyō imin" [Japanese fishing immigrants in Taiwan before World War II], *Tokyo Keidai Kaishi* 245 (2005): 98–100.

19. Hayase Shinzō, "Meiji-ki Manira-wan no Nihonjin gyomin" [Japanese fishermen in Manila Bay during the Meiji period], in *Kaijin no sekai* [The world of sea people], ed. Akimichi Tomoya (Tokyo: Dōbunkan Shuppan, 1998), 353–365.

20. Interview with Munaoka Tomio and Munaoka Chizuko, January 28, 2009; Kanezaki Gyogyō-shi Hensan Iinkai, ed., *Chikuzen Kanezaki gyogyō-shi* [The History of Kanezaki Fishery in Chikuzen] (Fukuoka: Kanezaki Gyogyō Kyōdō Kumiai, 1992), 908.

21. Nōji, Futamado, and Yoshiwa in Hiroshima Prefecture were famous as hometowns for the *ebune* fleet. The islands of Shimokamagari and Toyoshima in Hiroshima Prefecture, and Hirado and the Nishisonogi Peninsula in Nagasaki Prefecture, also hosted the *ebune* boats.

22. Byungchul Kim, "Sea Nomads of Japan," *International Journal of Maritime History* 11, no. 2 (1999): 87–105.

23. Mori Kōichi, "Kaijin bunka no butai" [The stage of sea people's culture], in *Ise to Kumano no umi* [The sea of Ise and Kumano], ed. Amino Yoshihiko et al. (Tokyo: Shōgakkan, 1992), 40.

24. Segawa Kiyoko, *Ama* [Women divers] (Tokyo: Miraisha, 1970), 177, 186.

25. Yoshida, *Chōsen suisan kaihatsu-shi,* 211.

26. For more on the historical background of *ama* divers of Kanezaki, see Kalland, *Fishing Villages in Tokugawa Japan,* chapter 10.

27. Interview with Munaoka Tomio and Munaoka Chizuko, January 28, 2009.

28. Segawa, *Ama,* 2.

29. Segawa Kiyoko, *Hisagime* [Female peddlers] (Tokyo: Miraisha, 1971), 73.

30. Jane Nadel-Klein and Dona Lee Davis, *To Work and to Weep: Women in Fishing Economies* (St. John's: Institute of Social and Economic Research, Memorial University of Newfoundland, 1988), 30.

31. Kawakami Masayuki, *Hiroshima Ōtagawa deruta no gyogyōshi* [The fisheries history of the Hiroshima Ōta River delta] (Hiroshima: Takumi Shuppan, 1976), 10.

32. Miyamoto Tsuneichi, *Suō-Ōshima o chūshin to shitaru umi no seikatsushi* [The record of sea life focusing on Suō-Ōshima] (Tokyo: Miraisha, 1994), 108; Nakamura Hiroko, "Hisagime: Gyōshō no hatten" [Female peddlers: The development of peddling], in *Kōza: Nihon no minzoku 5 seigyō* [Lectures on Japanese folklore, vol. 5, Occupation], ed. Kawaoka Takeharu (Tokyo: Yūseidō, 1980), 182.

33. Segawa, *Hisagime,* 111.

34. Takeda, *Manira e watatta setouchi gyomin,* 277, 310, 321; Hiroshimaken, ed., *Hiroshimaken-shi minzoku-hen* [The history of Hiroshima-ken, folklore] (Hiroshimashi: Hiroshimaken, 1978), 749.

35. Kawaoka, *Umi-no-tami,* 61–62.

36. Gotō Akira, *Umi no bunka-shi* [The cultural history of the sea] (Tokyo: Miraisha, 1996), 194.

37. Clear reasons the Nonami went there to start diving, and how, remain unknown.

38. David C. S. Sissons, "1871–1946 nen no Ōsutoraria no Nihonjin" [The Japanese in Australia between 1871 and 1946], *Ijūkenkyū* 10 (1974): 28–31.

39. Ogawa Taira, *Arafurakai no shinju* [Pearls of Arafura Sea] (Tokyo: Ayumi Shuppan, 1976), 17.

40. Shimizu Akira, ed., *Kinan no hitobito no kaigai taiken kiroku* 3 [The recorded experiences of Kinan people in a foreign land, vol. 3] (privately printed, 1993), 6.

41. "Nakasuji Gorokichi-rō (2)" [Mr. Nakasuji Gorokichi (2)], *Nippu Jiji,* April 22, 1929, 6.

Chapter 2: Japanese Fishermen Enter Hawaiian Waters

1. Gerald Horne, *The White Pacific: U.S. Imperialism and Black Slavery in the South Seas after the Civil War* (Honolulu: University of Hawai'i Press, 2007), 113.

2. Christofer H. Boggs and Bert S. Kikkawa, "The Development and Decline of Hawaii's Skipjack Tuna Fishery," *Marine Fisheries Review* 55, no. 2 (1993): 62.

3. Nippu Jijisha, *Kanyaku imin Hawai tokō gojusshūnen kinenshi* [The special volume commemorating the fiftieth anniversary of the arrival of the government contract immigrants] (Honolulu: Nippu Jijisha, 1935), 67.

4. Ibid., 19.

5. Wakayamaken, ed., *Wakayamaken iminshi* [The immigration history of Wakayama Prefecture] (Wakayamashi: Wakayamaken Prefectural Government, 1957), 511.

6. "Nakasuji Gorokichi-rō (2)" [Mr. Nakasuji Gorokichi (2)], *Nippu Jiji,* April 22, 1929, 6; "Nakasuji Gorokichi-rō (3)" [Mr. Nakasuji Gorokichi (3)], *Nippu Jiji,* April 24, 1929, 7.

7. Moke Manu, et al., *Hawaiian Fishing Traditions* (Honolulu: Kalamakū Press, 2006), 8–9.

8. Aubrey Haan and Albert L. Tester, "Hawai'i's Fishing Industry," *Hawaii Educational Review* 38, no. 3 (1949): 61. Later, the Japanese went into pond fishing. See Nippu Jijisha, *Nippujiji Hawai nenkan* [Nippu Jiji's almanac of the Japanese in Hawai'i] (Honolulu: Nippu Jijisha, 1928), 97.

9. Edward Glazier, Janna Shackeroff, Courtney Carothers, Julia Stevens, and Russell Scalf, *A Report on Historic and Contemporary Patterns of Change in Hawai'i-Based Pelagic Handline Fishing Operations—Final Report* (Honolulu: School of Ocean and Earth Science and Technology, University of Hawai'i at Mānoa, 2009), 8.

10. It is a "trolling hook made to skim over the surface of the water. The hook is made of pearl shell or bone, and the lure itself is also made of pearl shell of various types, each having a different name." Daniel Kahā'ulelio, *Ka'Oihana Lawai'a*: *Hawaiian Fishing Traditions* (Honolulu: Bishop Museum Press, 2006), 36.

11. Ibid., 25–29.

12. Margaret Titcomb, *Native Use of Fish in Hawaii* (Honolulu: University of Hawai'i Press, 1972), 13–14.

13. Nōshōmushō Suisankyoku, ed., *Nihon suisan hosaishi* [The record of Japanese fishing] (1911–1935; reprint, Tokyo: Iwasaki Bijutsusha, 1983), 173.

14. "Nakasuji Gorokichi-rō (4)," [Mr. Nakasuji Gorokichi (4)], *Nippu Jiji,* April 25, 1929, 7.

15. Manu et al., *Hawaiian Fishing Traditions,* xvi–xvii.

16. Ageno Kanzaburō, "Yonjū-gonen mae no gyogyō-kai" [The fisheries industry forty years ago], *Hawaii Times 60th Anniversary,* no. 2 (October 1, 1955): 12.

17. Hawai Nihonjin Iminshi Kankō Iinkai, ed., *Hawai Nihonjin iminshi* [A history of Japanese immigrants in Hawai'i] (Honolulu: United Japanese Society of Hawaii, 1962), 311–313.

18. Oliver P. Jenkins, *Report on Collection of Fishes Made in the Hawaiian Island, with Description of New Species* (Washington, DC: Government Printing Office, 1903), 419.

19. John R. K. Clark, *Guardian of the Sea: Jizo in Hawai'i* (Honolulu: University of Hawai'i Press, 2007), 6.

20. Gotō Akira, "Hawai Nikkei imin no gyogō to Nanki-chihō no kenken gyohō" [The fishing gear of Japanese immigrants in Hawai'i and kenken fishing in the Nanki area], *Mingu Kenkyū* 84 (1989): 5–6.

21. "Nakasuji Gorokichi-rō (5)" [Mr. Nakasuji Gorokichi (5)], *Nippu Jiji*, April 26, 1929, 8.

22. Wakayamaken,*Wakayamaken iminshi*, 516.

23. Kushimotochō-shi Hensan Iinkai, ed., *Kushimotochō-shi* [The history of Kushimotochō] (Kushimotochō: Kushimotochō Yakuba, 1995), 698.

24. Gotō, "Hawai Nikkei imin no gyogō to Nanki-chihō no kenken gyohō," 5.

25. John N. Cobb, "The Commercial Fisheries of the Hawaiian Islands in 1903," Department of Commerce and Labor, Bureau of Fisheries (Washington, DC: Government Printing Office, 1905), 507.

26. Ibid., 483, 484, 492, 495.

27. Shimizu Akira, ed., *Kinan no hitobito no kaigai taiken kiroku 1* [The recorded experiences of Kinan people in a foreign land, vol. 1] (privately printed, 1993), 16–17.

28. Shimizu Akira, ed., *Kinan no hitobito no kaigai taiken kiroku 3* [The recorded experiences of Kinan people in a foreign land, vol. 3] (privately printed, 1993), 23.

29. Kushimotochō-shi Hensan Iinkai, *Kushimotochō-shi*, 633.

30. Shimizu, *Kinan no hitobito no kaigai taiken kiroku 1*, 16–19.

31. Shimizu Akira, ed., *Kinan no hitobito no kaigai taiken kiroku 2* [The recorded experiences of Kinan people in a foreign land, vol. 2] (privately printed, 1993), 29; Kushimotochō-shi Hensan Iinkai, *Kushimotochō-shi*, 686.

32. "Hawai no katsuo-ryō wa Kishū shusshinsha no te ni yoru" [Skipjack tuna fishing has been conducted exclusively by Kishū fishermen], *Nippu Jiji*, January 10, 1907, 3.

33. Morimoto Takashi, *Tōwachōshi 3* [The History of Tōwachō, vol. 3] (Suō-Ōshimachō: Tōwachō Yakuba, 1986), 183–218.

34. "Konogoro no Hawai no gyogyō no yōsu" [Recent fishing enterprises in Hawai'i], *Nippu Jiji*, November 14, 1908, 4; November 23, 1908, 4.

35. Ibid.

36. Data are from Hakuseiji Temple, ed., *Kamuro-fukkokuban 1* [Kamuro reprinted version, vol. 1] (Kobe: Mizunowa Shuppan, 2001); *Kamuro-fukkokuban 2* [Kamuro reprinted version, vol. 2] (Kobe: Mizunowa Shuppan, 2002); *Kamuro-fukkokuban 3* [Kamuro reprinted version, vol. 3] (Kobe: Mizunowa Shuppan, 2002).

37. Ōkubo Gen'ichi, ed., *Hawai Nihonjin hatten meikan bōchōban* [The biographical record of the development of the Japanese in Hawai'i, Bōchō edition] (Honolulu: Hawai Shōgyōsha, 1940), 2.

38. Kitagawa Isojirō, "Hawai shima no gaikyō" [The situation in the island of Hawai'i], *Kamuro* 3 (May 5, 1915): 11–12.

39. Morita Sakae, *Hawai Nihonjin hatten-shi* [The development of Japanese history in Hawai'i] (Waipahu: Shin'eikan, 1915), 278; Hakuseiji Temple, *Kamuro fukkokuban 1*, 114.

40. In 1914, the Hilo Suisan started a fish auction.

41. Kitagawa, "Hawai shima no gaikyō," 10–12.

42. Hiroshimashi, ed., *Shinshū Hiroshimashi daisankan shakaikeizai-hen* [The new history of Hiroshima city, vol. 3, Society and economy] (Hiroshimashi: Hiroshima Shiyakusho, 1959), 520.

43. Sōga Yasutarō, *Gojūnenkan no Hawai kaiko* [Memoirs of fifty years in Hawai'i] (Honolulu: Gojūnenkan no Hawai Kaiko Kankōkai, 1953), 367.
44. Jenkins, *Report on Collections of Fishes Made in the Hawaiian Islands,* 419.
45. Chung Kun Ai, *My Seventy-Nine Years in Hawaii* (Hong Kong: Cosmorama Pictorial Publisher, 1960), 159–162.
46. Sōga, *Gojūnenkan no Hawai kaiko,* 367.
47. Michael M. Okihiro and Friends of A'ala, *A'ala: The Story of a Japanese Community in Hawaii* (Honolulu: Japanese Cultural Center of Hawai'i, 2003), 47.
48. Sōga, *Gojūnenkan no Hawai kaiko,* 368; Morita, *Hawai Nihonjin hatten-shi,* 277.
49. Morita, *Hawai Nihonjin hatten-shi,* 277–278.
50. "Imin Hyakunen" [One hundred years of immigration], *Hawaii Times,* September 22, 1966, 6.
51. Morita, *Hawai Nihonjin hatten-shi,* 277.
52. Kida Katsukichi, "Waga chichi o kataru" [Talking about my father], in *Fukkatsu jūgoshūnen kinenshi* [The fifteenth anniversary of the revival] (Honolulu: Hawai Wakayama Kenjinkai, 1963), 198.
53. Hawai Shinpōsha, *Hawai Nihonjin nenkan* [The almanac of the Japanese in Hawai'i] (Honolulu: Hawai Shinpōsha, 1921), 97–98.
54. Owen K. Konishi, "Fishing Industry of Hawaii with Special Reference to Labor" (Honolulu: University of Hawai'i Reports of Students in Economics and Business, 1930), 28–29; Donald M. Schug, "Hawaii's Commercial Fishing Industry: 1820–1945," *Hawaiian Journal of History* 35 (2001): 23.
55. The word might have originated in South China or Southeast Asia.
56. Konishi, "Fishing Industry of Hawaii with Special Reference to Labor," 32.
57. Kawaoka Takeharu, *Umi-no-tami* [The people of the sea] (Tokyo: Heibonsha, 1987), 200.
58. Konishi, "Fishing Industry of Hawai'i with Special Reference to Labor," 32.
59. Hawai Nihonjin Iminshi Kankō Iinkai, *Hawai Nihonjin iminshi,* 315–316.
60. Shimizu, *Kinan no hitobito no kaigai taiken kiroku 1,* 20.
61. "Imin Hyakunen," 6.
62. Okihiro and Friends of A'ala, *A'ala,* 25.
63. Susan Blackmore Peterson, "Decisions in a Market: A Study of the Honolulu Fish Auction" (PhD diss., University of Hawai'i, 1973), 126.
64. S. B. Dole, "The Old Fish Market," *Twenty-Ninth Annual Report of the Hawaiian Historical Society* (1921): 20.
65. "Hiro no hōjin gyogyōsha no toru sakana ga eiseihō ihan" [Japanese fishermen in Hilo violate sanitary regulations], *Nippu Jiji,* January 3, 1908, 5.
66. "Gyogyō tai shokuryō kyōkyū no mondai ni taishi" [Problems with fishing and the Department of Food], *Hawai Hōchi,* July 17, 1918, 5.
67. Earnest Wakukawa, *A History of the Japanese People in Hawaii* (Honolulu: Tōyō Shoin, 1938), 214.
68. Tin-Yuke Char, ed., *The Sandalwood Mountains: Readings and Stories of the Early Chinese in Hawaii* (Honolulu: University of Hawai'i Press, 1975).
69. "Boikotto nao keizoku seri" [The boycott is still going on], *Nippu Jiji,* February 5, 1910, 1.
70. "Nicchū nakagainin to gyogyō-gaisha no atsureki" [The friction between the Japanese and Chinese dealers and the fishing company], *Nippu Jiji,* February 4, 1910, 1.

71. "Nicchū gōdō nakagainin boikotto jiken" [The joint boycott of the Japanese and Chinese dealers], *Nippu Jiji,* February 9, 1910, 4.

72. Dennis M. Ogawa, *Kodomo no tameni: For the Sake of the Children* (Honolulu: University of Hawai'i Press, 1978), 133.

73. "Nicchū nakagainin to gyogyō-gaisha no atsureki," 1.

74. "Konogoro no Hawai no gyogyō no yōsu" [The fishing business in today's Hawai'i], *Nippu Jiji,* November 24, 1908, 4.

75. "The Japanese Tuna Fishermen of Hawaii," *Hawai Hōchi,* June 5, 1973, 1.

76. Hawai Nihonjin Iminshi Kankō Iinkai, *Hawai Nihonjin iminshi,* 27.

77. Nippu Jiji, *Hawai dōhō hatten kaikoshi* [Memoir of the development of Japanese in Hawai'i] (Honolulu: Nippu Jijisha, 1921), 60; Hawai Shinpōsha, *Hawai Nihonjin nenkan,* 132–138.

78. "Hawai enkai no gyogyōken kan'yū e" [Government ownership of Hawaiian waters], *Yamato Shinbun,* November 17, 1902, 1.

79. Jenkins, *Report on Collections of Fishes Made in the Hawaiian Islands,* 419. *Konohiki* rights, a relic of the ancient private ownership of ocean waters extending to the boundary reefs by the landowners of adjacent property, continued for decades after the annexation. The *konohiki,* or landowner, also owned the ocean bottom and the fish that frequented or inhabited such waters. See Frank T. Bell and Elmer Higgins, "A Plan for the Development of the Hawaiian Fisheries" (Washington, DC: US Government Printing Office, 1939), 19.

80. "No Anti-Japan Bills Are Wanted," *Pacific Commercial Advertiser,* February 20, 1909, 1–2.

81. Wakayamaken, *Wakayamaken iminshi,* 381–386; Sandy Lydon, *The Japanese in the Monterey Bay Region: A Brief History* (Capitola, CA: Capitola, 1997), 54–80.

82. Yoneyama Hiroshi, "Amerika-shi jojyutsu no ekkyō-ka to Nihonjin no kokusai idō: Iminshi no wakugumi no kaitai to saikōchiku ni mukete" [The transnationalization of the description of American history and the transnational movements of the Japanese: Toward the destruction and reconstruction of the contours of immigration history], *Ritsumeikan Bungaku* 597 (February 2007): 148–149. Andrew F. Smith revealed that the tuna canneries on Terminal Island constructed barracks and apartments and rented the apartments to Japanese fishermen, whom they deemed the "best fishermen." Canneries also offered bonuses and advanced money to Japanese fishermen to build or buy boats as inducements to sign contracts to fish for the company. Andrew F. Smith, *American Tuna: The Rise and Fall of an Improbable Food* (Berkeley: University of California Press, 2012), 50–51.

83. Yasutarō Sōga said the territorial government authorities attempted to curb the activities of the Japanese by importing a hundred Italians from the continent, but, for some reason, they did not stay long in Hawai'i and returned to the mainland. Thus far, no detailed records have been found to support his argument. Sōga, *Gojūnenkan no Hawai kaiko,* 367.

84. "Shin gyogyōhō jisshi ni tsuki" [On the enforcement of new fisheries law], *Hawai Shokumin Shinbun,* May 14, 1909, 2–3.

85. "Hilo Waiakea Suisan Kabushiki Gaisha funsō rakuchaku no tenmatsu" [The details of the dispute at the Hilo Waiakea Suisan Co.], *Hawai Shokumin Shinbun,* October 4, 1909, 5; October 6, 1909, 5; October 8, 1909, 5; October 11, 1909, 5; October 13, 5; October 15, 1909, 5; October 18, 1909, 5; October 20, 1909, 5.

86. "Futatabi nefu mondai ni tsuite" [The *nehu* problem again], *Hawai Hōchi,* September 8, 1913, 4.

87. "Sengyo gyōshōsha nomi ni eigyōzei ga menjo sareta wake" [The reason fish dealers are exempted from sales tax], *Nippu Jiji,* October 18, 1950, 7.

88. Ōtani Matsujirō, *Waga hito to narishi ashiato: Hachijū nen no kaiko* [Becoming a man: Memoirs of my eighty years] (Honolulu: M. Ōtani, 1971), 37–38.

89. Onodera Tokuji, ed., *Honolulu Nihonjin Shōgyō Kaigisho nenpō* [The annual report of the Japanese Chamber of Commerce in Honolulu] (Honolulu: Honolulu Nihonjin Shōgyō Kaigisho, 1922), 149.

Chapter 3: The Heyday of the Japanese Fishing Industry in Hawaiʻi

1. Nōshōmushō Suisankyoku, *Kaigai ni okeru honpō hōjin no gyogyō jōkyō* [The overseas fishing of the Japanese] (Tokyo: Nōshōmushō Suisankyoku, 1918), 176.

2. Hisao Goto, Kazuo Shinoto, and Alexander Spoehr, "Craft History and the Merging of Tool Traditions: Carpenters of Japanese Ancestry in Hawaii," *Hawaiian Journal of History* 17 (1983): 158, 169.

3. Ibid., 158, 168–170.

4. Hans Konrad Van Tilburg, "Vessels of Exchange: The Global Shipwright in the Pacific," in *Seascapes: Maritime Histories, Littoral Cultures, and Transoceanic Exchanges,* ed. Jerry H. Bentley, Renate Bridenthal, and Kären Wigen (Honolulu: University of Hawaiʻi Press, 2007), 47.

5. Ōtani Matsujirō, *Waga hito to narishi ashiato: Hachijū nen no kaiko* [Becoming a man: Memoirs of my eighty years] (Honolulu: M. Ōtani, 1971), 48.

6. Nippu Jijisha, *Nippujiji Hawai nenkan* [Nippu Jiji's almanac of the Japanese in Hawaiʻi] (Honolulu: Nippu Jijisha, 1929), 104. When the Japanese travel somewhere, many of them buy gifts for their families, friends, neighbors, and colleagues. This custom originates from pilgrimages to famous shrines and temples in the premodern period. Prior to departure, the pilgrims received money and gifts for travel from the people of their communities. As evidence that the pilgrims visited the designated places, they purchased religious charms or local specialties and gave such items to those who invested in them as *omiyage*. Such monetary and material exchanges remain even in today's Japan.

7. "Nichibeijin gōdō de kanzume kaisha setsuritsu" [The establishment of canneries by Japanese and Americans], *Nippu Jiji,* September 26, 1922, 3; Nippu Jijisha, *Nippu Jiji Hawai nenkan* [Nippu Jiji's almanac of the Japanese in Hawaiʻi] (Honolulu: Nippu Jijisha, 1931), 124.

8. It was, however, unable to make a go of it, because, according to Kamezō Matsuno, it had "marketing problems." It ceased to operate in 1938, when local organizations, led by the Hilo Woman's Club, complained that "it would stink up the town." Since then, Suisan Co. has sold skipjack tuna to Hawaiian Tuna Packers. "Fishing Firm Founder Is Still Active at 77," *Honolulu Advertiser,* October 30, 1955, A7.

9. "The Japanese Tuna Fishermen of Hawaii," *Hawai Hōchi,* June 5, 1973, 1.

10. Interview with Shimizu Hisao and Shimizu Shizue, March 3, 2008.

11. Wakayamaken, ed., *Wakayamaken iminshi* [The immigration history of Wakayama Prefecture] (Wakayamashi: Wakayamaken, 1957), 512–513.

12. Nippu Jijisha, *Nippujiji Hawai nenkan* [Nippu Jiji's almanac of the Japanese in Hawaiʻi] (Honolulu: Nippu Jijisha, 1928), 98–100.

13. Owen K. Konishi, "Fishing Industry of Hawaii with Special Reference to Labor" (Honolulu: University of Hawaiʻi Reports of Students in Economics and Business, 1930), 38; interview with Akira Ōtani, September 4, 2007.

14. Mike Markrich, "Fishing for Life," in *Kanyaku Imin: A Hundred Years of Japanese Life in Hawaii*, ed. Leonard Lueras (Honolulu: International Savings and Loan Association, 1985), 142; interview with Akira Ōtani, March 3, 2008.

15. Interview with Donald Kida, September 9, 2008.

16. Markrich, "Fishing for Life," 142.

17. Ōtani Matsujirō, "Nikkei gyogyō kaisha no hensen o kataru" [Talking about the development of the Japanese fishing companies], *Hawaii Times 60th Anniversary*, no. 9 (October 1, 1955), 10.

18. Interview with Teruo Funai, March 3, 2008.

19. Susamichōshi Hensan Iinkai, ed., *Susamichōshi 2* [The history of Susamichō, no. 2] (Susamichō, Wakayama: Susamichō, 1978), 294.

20. Interview with Teruo Funai, March 3, 2008.

21. Susamichōshi Hensan Iinkai, *Susamichōshi 2*, 294.

22. John R. K. Clark, *Guardian of the Sea: Jizo in Hawai'i* (Honolulu: University of Hawai'i Press), 6.

23. Hawai Shinpōsha, *Hawai Nihonjin nenkan* [Almanac of the Japanese in Hawai'i] (Honolulu: Hawai Shinpōsha, 1924), 132–138; Nippu Jijisha, *Nippujiji Hawai nenkan* [Nippu Jiji's almanac of the Japanese in Hawai'i] (Honolulu: Nippu Jijisha, 1931–1932), 98–100.

24. Goto, Shinoto, and Spoehr, "Craft History and the Merging of Tool Traditions," 169.

25. H. Hamamoto, "The Fishing Industry of Hawaii" (BA thesis, University of Hawai'i, 1928), 25; Jack Y. Tasaka, "Hawai to Wakayama kenjin" [Hawai'i and people from Wakayama Prefecture], *Journal of the Pacific Society* 31 (July 1986): 66.

26. "The Japanese Tuna Fishermen of Hawaii," 1.

27. Hamamoto, "The Fishing Industry of Hawaii," 26–28; Konishi, "Fishing Industry of Hawaii with Special Reference to Labor," 17–18.

28. Hamamoto, "The Fishing Industry of Hawaii," 26–28; Konishi, "Fishing Industry of Hawaii with Special Reference to Labor," 17–18; interview with Teruo Funai, March 2008.

29. Linda Lucas Hudgins and Samuel G. Pooley, "Growth and Contraction of Domestic Fisheries: Hawaii's Tuna Industry in the 1980s," in *Tuna Issues and Perspectives in the Pacific Islands Region*, ed. David J. Doulman (Honolulu: East-West Center, 1987), 226.

30. Hamamoto, "The Fishing Industry of Hawaii," 32–33; Daniel Kahā'ulelio, *Ka'Oihana Lawai'a: Hawaiian Fishing Traditions* (Honolulu: Bishop Museum Press, 2006), 99.

31. Hamamoto, "The Fishing Industry of Hawaii," 35–36.

32. Hudgins and Pooley, "Growth and Contraction of Domestic Fisheries," 230. Japanese fishermen formed a large fishing community on Terminal Island in Los Angeles, and predominantly engaged in tuna fishing. They developed their own fishing styles that were not always identical to those adopted in Hawai'i. For detailed information on fishing styles and gear used on Terminal Island, see Andrew F. Smith, *American Tuna: The Rise and Fall of an Improbable Food* (Berkeley: University of California Press, 2012), 52–54.

33. Hamamoto, "The Fishing Industry of Hawaii," 29–32; Susan Blackmore Peterson, "Decisions in a Market: A Study of the Honolulu Fish Auction" (PhD diss., University of Hawai'i, 1973), 41.

34. Interview with Akira Ōtani, March 3, 2008.
35. Interview with Teruo Funai, March 3, 2008.
36. Hawai Shinpōsha, *Hawai Nihonjin nenkan* [Almanac of the Japanese in Hawai'i] (Honolulu: Hawai Shinpōsha, 1921): 132–138; Iida Kōjirō, *Hawai Nikkeijin no rekishi chiri* [The history and geography of Japanese Americans in Hawai'i] (Kyoto: Nakanishiya Shuppan, 2003), 76–78, 103.
37. Interview with Hisao Shimizu and Shizue Shimizu, March 3, 2008.
38. Ibid.
39. "Farmers of the Sea," *Sales Builder* 5, no. 9 (September 1934): 5.
40. Chris Friday, *Organizing Asian American Labor: The Pacific Coast Canned-Salmon Industry, 1870–1942* (Philadelphia: Temple University Press, 1994), 43, 116–118.
41. Daphne Marlatt, *Steventon Recollected: Japanese-Canadian History* (Vancouver: Provincial Archives of British Columbia, 1975), 23.
42. Friday, *Organizing Asian American Labor,* 117–118; interview with Nancy Ōtani and Evelyn Ōtani, September 4, 2007; interview with Shizue Shimizu, March 3, 2008.
43. Ethel Erickson, "Earnings and Hours in Hawaii Woman-Employing Industries," Bulletin of the Women's Bureau (Washington, DC: US Government Printing Office, 1940), 19. Data in this report reveal that about 87 percent of female workers were Japanese, but it includes two canneries, a can factory, a small bakery, and two cotton mattress factories. Since the tuna canneries and the can factory outstripped the others in number of employees, it indicates Japanese women's dominance in the tuna canneries.
44. Ethnic Studies Oral History Project, Ethnic Studies Program, "Tsuru Yamauchi," in *Uchinanchu: A History of Okinawans in Hawaii* (Honolulu: University of Hawai'i at Manoa, 1981), 501–502.
45. Ibid., 502.
46. Interview with Shizue Shimizu, March 3, 2008.
47. Erickson, "Earnings and Hours in Hawaii Woman-Employing Industries," 3.
48. Interview with Teruo Funai, March 3, 2008.
49. Ōtani, *Waga hito to narishi ashiato,* 34.
50. Interview with Akira Ōtani, March 3, 2008.
51. Michi Kodama-Nishimoto, Warren S. Nishimoto, and Cynthia A. Oshiro, eds., *Hanahana: An Oral History Anthology of Hawaii's Working People,* Ethnic Studies Oral History Project (Honolulu: University of Hawai'i at Manoa, 1984), 80–81.
52. Theodore C. Bestor, *Tsukiji: The Fish Market at the Center of the World* (Berkeley: University of California Press, 2004), 83–84.
53. "Ōtani Matsujirō-den" [A life of Matsujirō Ōtani], *Nippu Jiji,* December 4, 1941, 6.
54. "Aala Section Has Become Market Place," *Honolulu Star-Bulletin,* December 4, 1941, 3; Ōtani, *Waga hito to narishi ashiato,* 53–58.
55. "Market Place Is Operated by Matsujiro Otani, Ltd.," *Honolulu Star-Bulletin,* December 4, 1941, 3.
56. Erickson, "Earnings and Hours in Hawaii Woman-Employing Industries," 2.
57. Peterson, "Decisions in a Market," 129.
58. "Gyofu no tsuma" [The wife of a fisherman], *Hawai Shokumin Shinbun,* July 15, 1910, 2.
59. Bestor, *Tsukiji,* 170.

60. "Konpira shaden kaichikuhi sangatsu yori boshū kaishi" [We accept donations for the repair of the Konpira Shrine from March], *Maui Shinbun,* February 25, 1941, 2.

61. Maeda Takakazu, *Hawai no jinjya-shi* [The history of Shinto shrines in Hawai'i] (Tokyo: Taimeidō, 1999), 206–210; "History of the Shrine," Hawaii Kotohira Jinsha-Hawaii Dazaifu Tenmangu, http://www.e-shrine.org/history.html, accessed April 28, 2014.

62. Clark, *Guardian of the Sea,* 50; "Kakaako chihō" [Kakaako area], *Nippu Jiji,* October 11, 1922, 5.

63. "Honolulu tsūshin" [News from Honolulu], *Kamuro* 82 (May 1929): 9; "Honolulu tsūshin," *Kamuro* 90 (March 1931): 6; "Honolulu tsūshin," *Kamuro* 91 (August 1931): 6; "Hawai Honolulu Hachiman-kō no kōen" [Support from Hachiman-kō in Honolulu, Hawai'i], *Kamuro* 140 (May 1938): 4.

64. "Kakaako chihō 4" [Kaka'ako area 4], *Nippu Jiji,* October 10, 1922, 5.

65. Gary Y. Okihiro, *Cane Fires: The Anti-Japanese Movement in Hawai'i, 1865–1945* (Philadelphia: Temple University Press, 1991), 70–76.

66. Of course, it is assumed that fishermen supported the strike by providing fresh fish and other marine products to the strikers and their families. Their involvement in the strike of 1920 should be examined further.

67. Around 1930, Hawaiian waters had 359 sampans equipped with gasoline- or diesel-powered engines. The largest ones were 135 horsepower; the smallest were 3 horsepower. Almost all the owners of these boats were Japanese. In Honolulu, Hawaii Suisan Co. had 62 affiliated sampans, Pacific Fishing Co. had 51, and Honolulu Fishing Co. had 37. Outside of Honolulu, 44 sampans operated independently on O'ahu. On the island of Hawai'i, Suisan Co. (Suisan Kabushiki Kaisha) had 54 affiliated sampans, and Hawaii Island Fishing Co. had 26. The islands of Kaua'i, Maui, Moloka'i, and Lāna'i did not have any Japanese fishing companies in those days; the Japanese sampans operated independently, although some of them formed cooperative organizations and helped each other. Kaua'i had 45 sampans, Maui 37, Moloka'i two, and Lāna'i had one sampan. Nippu Jijisha, *Nippujiji Hawai nenkan* (1928), 101–104; Nippu Jijisha, *Nippujiji Hawai nenkan* (1930), 115–119.

68. Nippu Jijisha, *Nippujiji Hawai nenkan* (1929), 106.

69. Okumura Takie, *Hawai ni okeru nichibei mondai kaiketsu undō* [The movement for the solution of Japan-US problems in Hawai'i] (Kyoto: Naigai Shuppan Insatsu, 1925), 65.

70. Dennis M. Ogawa, *Kodomo no tame ni: For the sake of the children* (Honolulu: University of Hawai'i Press, 1978), 198–199.

71. Nippu Jijisha, *Nippujiji Hawai nenkan* (1931), 125.

72. Ageno Kanzaburō, "Hawai gyogyōkai ni sasagu" [I dedicate myself to the fisheries in Hawai'i], *Nippu Jiji,* August 15, 1941, 2.

73. Ibid.

74. Ageno Kanzaburō, "Hawai no gyogyō o ronzu" [Talking about commercial fishing in Hawai'i], in *Hawai Nihonjin jitsugyō shōkaishi* [The introduction of Japanese industries in Hawai'i], ed. Hawai Hōchi (Honolulu: Hawai Hōchi, 1941), 44.

75. Matsumoto Sei, "Hawai no shokuryō mondai to suisan gakkō no setsuritsu" [The food problem in Hawai'i and establishment of a fisheries school], in *Hawai Nihonjin jitsugyō shōkaishi,* 47.

76. Nippu Jijisha, *Nippujiji Hawai nenkan* (1932), 100.

77. Okihiro, *Cane Fires*, 157.
78. "Young Generation and Fishing," *Nippu Jiji*, April 9, 1929, 2.
79. Nippu Jijisha, *Nippujiji Hawai nenkan* (1937), 124.
80. Interview with Brooks Takenaka, September 4, 2008.
81. Nippu Jijisha, *Nippujiji Hawai nenkan* (1940), 120; "Hawai no gyogyō" [Fisheries in Hawai'i], in *Hawai Nihonjin jitsugyō shōkaishi*, 40.
82. The issei Japanese were not permitted to become naturalized American citizens until 1952.
83. Oliver P. Jenkins, "Report on Collections of Fishes Made in the Hawaiian Islands, with Descriptions of New Species," Series US Commission of Fish and Fisheries, Doc. 534 (Washington, DC: Government Printing Office, 1903), 419.
84. "Hawai gyogyō no shōrai" [The future of fisheries in Hawai'i], *Hawai Shokumin Shinbun*, October 5, 1910, 5.
85. Ōtani Matsujirō, "Nikkei gyogyō kaisha no hensen o kataru," 10.
86. Nippu Jijisha, *Nippujiji Hawai nenkan* (1928), 98; ibid. (1932–1933), 101–102; ibid. (1935–1936), 109; ibid. (1936–1937), 114–115; ibid. (1937–1938), 124.

Chapter 4: Surviving the Dark Days

1. Gary Y. Okihiro, *Cane Fires: The Anti-Japanese Movement in Hawai'i, 1865–1945* (Philadelphia: Temple University Press, 1991), 96, 110, 112–113.
2. "Nihonjin gyosen 400 seki" [Four hundred Japanese fishing boats], *Nippu Jiji*, May 19, 1930, 3.
3. "Kaiketsu hōhō ari hikan wa muyō" [There are solutions; don't be pessimistic], *Nippu Jiji*, May 20, 1930, 3.
4. "Zeikan daihyō-sha no kondan" [A talk with customs representatives], *Nippu Jiji*, May 27, 1930, 3.
5. "Honeori-zon no kutabire mouke" [A lot of trouble for nothing], *Nippu Jiji*, June 30, 1930, 3.
6. "Hausuton daigishi e arigatō no denpō" [A telegram of thanks to Delegate Houston], *Nippu Jiji*, June 30, 1930, 2.
7. "Gyokakubutsu kazei ni taishi Hiro no yoron" [Public opinion in Hilo on the taxation of fish], *Nippu Jiji*, May 26, 1930, 3.
8. Hawai Shinpōsha, *Hawai Nihonjin nenkan* [Almanac of the Japanese in Hawai'i] (Honolulu: Hawai Shinpōsha, 1933–1934), 96.
9. Letter from N. B. Csofield to H. L. Kelly, May 8, 1935, Hawai'i State Archives; letter from Fred Schilling to I. H. Wilson, December 16, 1938, Hawai'i State Archives.
10. Letter from director, Hawaiian Tuna Packers, Ltd., to Admiral H. R. Yarnell, July 1, 1935, Hawai'i State Archives.
11. Ibid.
12. Hawai Shinpōsha, *Hawai Nihonjin nenkan* [Almanac of the Japanese in Hawai'i] (Honolulu: Hawai Shinpōsha, 1931–1932), 124.
13. Sayuri Guthrie-Shimizu, "Occupation Policy and the Japanese Fisheries Management Regime, 1945–1952," in *Occupied Japan: The U.S. Occupation and Japanese Politics and Society*, ed. Mark E. Caprio and Yoneyuki Sugita (New York: Routledge, 2007), 48–53.
14. Hans Konrad Van Tilburg, "Vessels of Exchange: The Global Shipwright in the Pacific," in *Seascapes: Maritime Histories, Littoral Cultures, and Transoceanic*

Exchanges, ed. Jerry H. Bentley, Renate Bridenthal, and Kären Wigen (Honolulu: University of Hawai'i Press, 2007), 44.

15. Interview with Shizue Shimizu, September 1, 2008.

16. Kida Katsukichi, "Waga chichi o kataru" [Talking about my father], in *Fukkatsu jūgoshūnen kinenshi* [The fifteenth anniversary of the revival] (Honolulu: Hawai Wakayama Kenjinkai, 1963), 199.

17. Nippu Jijisha, *Nippujiji Hawai nenkan* [Nippu Jiji's almanac of the Japanese in Hawai'i] (Honolulu: Nippu Jijisha, 1928), 100; Nippu Jijisha, *Nippujiji Hawai nenkan* [Nippu Jiji's almanac of the Japanese in Hawai'i] (Honolulu: Nippu Jijisha, 1940), 108.

18. "U.S. Charges Huge Sampan Conspiracy Here," *Honolulu Advertiser,* March 1, 1941, 3.

19. "Gyosen jiken" [The case of fishing boats], *Nippu Jiji,* March 18, 1941, 3; "Gyosen jiken" [The case of fishing boats], *Nippu Jiji,* March 19, 1941, 3.

20. "Gyosen bosshū saiban" [The trial of an impounded fishing boat], *Nippu Jiji,* June 10, 1941, 7.

21. "Bosshū hanketsu no nana gyosen ni tōkyoku no kandai sochi" [The authorities made a lenient decision for seven confiscated fishing boats], *Nippu Jiji,* August 27, 1941, 6.

22. "Gyoka neagari bōshi ni sangaisha-gawa de doryoku" [Three fishing companies were making efforts to avoid a rise in the price of fish], *Nippu Jiji,* March 5, 1941, 5.

23. "U.S. Charges Huge Sampan Conspiracy Here," 3.

24. "Dōhō gyogyōka o chūshō no Taizā-shi kokuhatsu sareru" [The Tiser was charged with defamation of Japanese fishermen], *Nippu Jiji,* March 4, 1941, 3.

25. Frank T. Bell and Elmer Higgins, "A Plan for the Development of the Hawaiian Fisheries," (Washington, DC: US Government Printing Office, 1939).

26. "Gyorui hogo zōshoku kenkyū-an" [A bill for the protection and increase of fish], *Nippu Jiji,* March 31, 1941, 2.

27. Susamichō-shi Hensan Iinkai, ed., *Susamichō-shi 2* (Susamichō, Wakayama: Susamichō, 1978), 289–290.

28. Hawai Wakayama Kenjinkai, ed., *Fukkatsu jūgoshūnen kinenshi* [The fifteenth anniversary of the revival] (Honolulu: Hawai Wakayama Kenjinkai, 1963), 74; Patsy Sumie Saiki, *Ganbare! An Example of Japanese Spirit* (Honolulu: Kisaku, 1982), 1–5.

29. Mike Markrich, "Fishing for Life," in *Kanyaku Imin: A Hundred Years of Japanese Life in Hawaii,* ed. Leonard Lueras (Honolulu: International Savings and Loan Association, 1985), 142.

30. T. T. Iwashita, "Development and Design of the Hawaiian Fishing Sampans," paper presented at the Hawaii Section of the Society of Naval Architects and Marine Engineers, January 1956, 5.

31. Van Tilburg, "Vessels of Exchange," 48.

32. Iwashita, "Development and Design of the Hawaiian Fishing Sampans," 5.

33. Jeffery F. Burton and Mary M. Farrell, *World War II Japanese American Internment Sites in Hawai'i* (Honolulu: Japanese Cultural Center of Hawai'i, 2007), 1.

34. Dennis M. Ogawa and Evarts C. Fox Jr., "Japanese Internment and Relocation: The Hawaii Experience," in *Japanese Americans: From Relocation to Redress,* ed. Roger Daniels, Sandra Taylor, and Harry H. L. Kitano (Seattle: University of Washington Press, 1986).

35. Susamichō-shi Hensan Iinkai, *Susamichō-shi 2*, 290.
36. Interview with Donald Kida, September 9, 2008.
37. Ōtani Matsujirō, *Waga hito to narishi ashiato: Hachijū nen no kaiko* [Becoming a man: Memoirs of my eighty years] (Honolulu: M. Ōtani, 1971), 58–62; interview with Akira Ōtani, March 3, 2008.
38. Ōtani, *Waga hito to narishi ashiato,* 63–84.
39. Interview with Shizue Shimizu, March 4, 2008.
40. Ibid.
41. Ibid.
42. Ibid.
43. Shimada Noriko, *Sensō to imin no shakaishi: Hawai nikkei Amerika-jin no Taiheiyō Sensō* [The social history of the war and the immigrants: The Pacific War of Japanese Americans in Hawai'i] (Tokyo: Gendai Shiryō Shuppan, 2004), 39–41.
44. Iwashita, "Development and Design of the Hawaiian Fishing Sampans," 5.
45. Memorandum, Territory of Hawai'i Office of the Military Governor, December 30, 1942, University of Hawai'i at Manoa, Hawaii War Records Depository (HWRD), Reel 5.
46. Letter from American Factors, Limited to Director of Food Control, April 17, 1942, HWRD, Reel 5.
47. Memorandum, Territory of Hawaii Office of the Military Governor, September 2, 1942, HWRD, Reel 5.
48. Memorandum from H. H. Warner to A. W. Macdonald, January 15, 1943, HWRD, Reel 16.
49. Letter from H. E. Raber to George Montgomery, October 16, 1943, HWRD, Reel 17.
50. Letter from A. H. Rice Jr., December 16, 1941, HWRD, Reel 5.
51. "Record of Production—December 7, 1941 to date," January 11, 1943, HWRD, Reel 5.
52. "Plan to Utilize Fishing Sampans Now Idle," March 23, 1942, HWRD, Reel 5.
53. Letter from Walter F. Dillingham to Frank H. West, November 6, 1942, HWRD, Reel 5.
54. Letter from F. H. West to W. F. Dillingham, November 18, 1942, HWRD, Reel 5.
55. "Memorandum for Files," November 23, 1942, HWRD, Reel 5; "Production," November 27, 1942, HWRD, Reel 5.
56. "Memorandum for Files," November 23, 1942, HWRD, Reel 5; "Production," November 27, 1942, HWRD, Reel 5. Later, Ernest Steiner of Hawaiian Tuna Packers took the post. Frank West volunteered his services, whereas Ernest Steiner, as of February 1943, received a monthly salary of $275.42. Memorandum from the Fishing Division to W. F. Dillingham, director of food production, February 9, 1943, HWRD, Reel 10.
57. Letter from F. W. Broadbent to Walter F. Dillingham, January 15, 1943, HWRD, Reel 5.
58. Letter from F. H. West to David Fleming, January 16, 1943, HWRD, Reel 5.
59. Memorandum from R. Penhallow to Dillingham, November 24, 1942, HWRD, Reel 5; "Permission for Pascal Nahoopii and Party to Fish in Shore Waters between Hauula and Kaluanui," November 13, 1942, HWRD, Reel 5.

60. Memorandum from W. F. Dillingham to Frank West, January 11, 1943, HWRD, Reel 5; Memorandum from the Fishing Division to W. F. Dillingham, February 9, 1943, HWRD, Reel 10.
61. Memorandum from H. H. Warner to A. W. Macdonald, January 15, 1943, HWRD, Reel 16.
62. Memorandum from the Fishing Division to W. F. Dillingham, February 9, 1943, HWRD, Reel 5.
63. Letter from W. F. Dillingham to Delos C. Emmons, January 13, 1943, HWRD, Reel 10.
64. Letter from David Fleming to F. H. West, January 30, 1943, HWRD, Reel 5.
65. Ibid.
66. The Federal Surplus Commodities Corporation (later known as the Commodity Credit Corporation) was a division of the War Food Administration. The FSCC was designated as the exclusive import trader supplying food to Hawai'i.
67. Letter from David Fleming to F. H. West, February 10, 1943, HWRD, Reel 5.
68. Iwashita, "Development and Design of the Hawaiian Fishing Sampans," 6.
69. Shimizu Akira, ed., *Kinan no hitobito no kaigai taiken kiroku 1* [The recorded experiences of Kinan people in a foreign land, vol. 1] (privately printed, 1993), 21.
70. Letter to the director of food control from Hawaiian Tuna Packers, May 5, 1944, HWRD, Reel 5.
71. "Suisan Begins Its Second Century," *Hawai'i Herald,* August 1, 2008, 6.
72. "The Japanese Tuna Fishermen of Hawaii," *Hawai Hōchi,* June 5, 1973, 1.
73. Ibid.

Chapter 5: The Reconstruction and Revitalization of Fisheries after World War II

1. Interview with Donald Kida, September 9, 2008.
2. Ibid.
3. Mike Markrich, "Fishing for Life," in *Kanyaku Imin: A Hundred Years of Japanese Life in Hawaii,* ed. Leonard Lueras (Honolulu: International Savings and Loan Association, 1985), 143.
4. "Nikkeijin no gyūjiru Hawai gyogyō" [Fishing industry in Hawai'i under Japanese control], *Hawaii Times*, October 28, 1950, 8.
5. P. V. Garrod and K. C. Chong, *The Fresh Fish Market in Hawaii* (Honolulu: Hawaii Agricultural Experiment Station, College of Tropical Agriculture, University of Hawai'i, June 1978), 11; interview with Akira Ōtani, September 4, 2007.
6. Arnold T. Hiura, "Suisan Begins Its Second Century," *Hawai'i Herald*, August 1, 2008, 7.
7. Kurt E. Kawamoto, Russell Y. Ito, Raymond P. Clarke, and Allison A. Chun, "Status of the Tuna Longline Fishery in Hawaii, 1987–88," Southwest Fisheries Center Administrative Report (Honolulu: National Marine Fisheries Service, 1989), 1.
8. "Honoruru maguro sen'in kumiai" [Honolulu Tuna Fishermen's Association], *Hawaii Times*, January 11, 1952, 3; "Katsuo sen sen'in kumiai" [Honolulu Skipjack Tuna Fishermen's Association], *Hawaii Times*, February 5, 1952, 4.
9. Interview with Akira Ōtani, September 4, 2007.
10. Ibid.

11. "Umi no wakoudo o kangei" [Welcome the youth of the sea], *Hawaii Times,* February 9, 1953, 3.

12. Interview with Hideo Tagawa, December 14, 2009.

13. Dennis M. Ogawa, *Kodomo No Tame Ni: For the Sake of the Children* (Honolulu: University of Hawai'i Press, 1978), 313–328.

14. Shimada Noriko, *Sensō to imin no shakaishi: Hawai nikkei Amerika-jin no Taiheiyō Sensō* [The social history of the war and the immigrants: The Pacific War of Japanese Americans in Hawai'i] (Tokyo: Gendaishiryō Shuppan, 2004), 14–15.

15. Maeda Takakazu, *Hawai no jinjya-shi* [The history of Shinto shrines in Hawai'i] (Tokyo: Taimeidō, 1999), 58–63.

16. Interview with Shizue Shimizu, March 3, 2008.

17. Donald E. Collins, "Wirin, Abraham Lincoln," in *Encyclopedia of Japanese American History,* ed. Brian Niiya (New York: Facts on File, 2001), 412–413.

18. Interview with Shizue Shimizu, March 3, 2008.

19. Ibid.; interview with Shizue Shimizu, September 1, 2008.

20. Interview with Hisao Shimizu, March 3, 2008.

21. US Bureau of Commercial Fisheries, *Hawaii Fishery Training Prospectus* (Honolulu: Hawaii Area Office, US Bureau of Commercial Fisheries, n.d.), 32.

22. Ōtani Matsujirō, *Waga hito to narishi ashiato: Hachijū nen no kaiko* [Becoming a man: Memoirs of my eighty years] (Honolulu: M. Ōtani, 1971), 118.

23. Vernon E. Brock, "A Proposed Program for Hawaiian Fisheries," *Hawaii Marine Laboratory Technical Report* no. 6 (February 1965): 3.

24. Ibid., 1.

25. Garrod and Chong, *The Fresh Fish Market in Hawaii,* 5.

26. Taiyo Gyogyo owned many cargo ships named *Banshū-maru* and distinguished each by number. The official name of this ship was *Dai-sanjū-go* [the thirty-fifth] *Banshū-maru.*

27. Ōtani, *Waga hito to narishi ashiato,* 142.

28. For the worldwide presence of Japanese fishing fleets during the 1950s, see Georg Borgstrom, *Japan's World Success in Fishing* (London: Fishing News, 1964).

Chapter 6: Okinawa and Hawai'i

1. Boys around the age of ten were recruited in exchange for a certain amount of money paid in advance to their parents. As novice fishermen at the bottom of the labor ladder, they underwent rigorous training in diving and fishing. With the improvement of their skills, their positions in the fleet rose. Their contracts usually expired before they reached twenty years of age, when they took a physical examination for conscription into the military. This system, called Ichuman-ui or Itoman-uri (selling off to Itoman), was outlawed as human trafficking in 1955. Thereafter, *agyā* netting substantially dwindled. Some scholars argue that the Itoman-uri contributed to solving overpopulation problems in poor villages and gave young men a chance to acquire fishing skills and become independent fishermen after the expiration of the contract. For the rise and fall of the system of Itoman-uri, see Ichikawa Hideo, *Itoman gyogyō no tenkai kōzō* [The development and structure of fishing in Itoman] (Naha: Okinawa Times, 2009), 63–77.

2. Ichikawa, *Itoman gyogyō no tenkai kōzō,* 95–97.

3. For the general history and culture of fishing in Okinawa and especially in Itoman, see Nakadate Kou, ed., *Nihon ni okeru kaiyōmin no sōgōteki kenkyū, jōkan* and

gekan [The general study of the sea people in Japan], 2 vols. (Fukuoka: Kyushu Daigaku Shuppankai, 1987, 1989); Ichikawa, *Itoman gyogyō no tenkai kōzō*.

4. For example, Kameho Uehara, a nisei whose parents were from Itoman, was shot to death on the day of the Pearl Harbor attack. He was strafed by a US combat plane while longline fishing for tuna. Itomanshi-shi Hensan Iinkai, ed., *Itomanshi-shi: Shiryō-hen 7, senjishiryō jōkan* [History of Itoman: Historical document 7, wartime records, vol. 1] (Itoman: Itoman Shiyakusho, 2003), 109.

5. Tomonori Ishikawa, "Historical Geography of Early Immigrants," in *Uchinanchu: A History of Okinawans in Hawaii,* ed. Ethnic Studies Oral History Project, Ethnic Studies Program (Honolulu: University of Hawai'i at Manoa, 1981), 96. Henry S. H. Yuen mentions that Okinawan fishermen are believed to have started the *ika-shibi* fishery as they had done in Okinawa, although the exact year they started using this method is unknown. *Ika* (squid)-*shibi* (yellowfin tuna) fishing uses handlines, squid-baited hooks, and a source of light to help attract baitfish, squid, and yellowfin tuna. Yuen, *A Night Handline Fishery for Tunas* (Southwest Fisheries Center Administrative Report, no. 2H, 1977), 1.

6. Henza Konjaku Shashinshū Henshū Iinkai, ed., *Hiyamuza Kanamori: Shashin ni miru Henza konjyaku* [The past and the present pictures of Henza] (Yonashiro: Henza Jichikai, 1984), especially 135–140, 164–174; Ushiban Seikichi, "Hawai imin" [Immigration to Hawai'i], in *Furuki o tazunete* [Cherishing of the old], ed. Henza Jichikai (Yonashiro: Henza Jichikai, 1985), 345–348; Seki Reiko, "Kaihatsu no umi ni shūsan suru hitobito: Henza ni okeru gyogyō no isou to mainā-sabushisutensu no tenkai" [The people who meet and part at the sea and development: The phase of fisheries on Henza and the development of minor subsistence], in *Okinawa rettō: Shima no shizen to dentō no yukue* [The Okinawa archipelago: Future of the islands' nature and tradition], ed. Matsui Ken (Tokyo: Tokyo Daigaku Shuppan, 2004): 146.

7. Such problems linger and are among the most sensitive problems for US-Japan relations. Many studies are available on the problems of the US military base in Okinawa. For general information published by the Military Base Affairs Section of Okinawa Prefecture, see its website, http://www.pref.okinawa.jp/kititaisaku/D-mokuji.html, accessed April 25, 2014; for sexual crimes committed by US servicemen, see Takazato Suzuyo, *Onna tachi no Okinawa* [Women's Okinawa] (Tokyo: Akashi Shoten, 1996).

8. Mitsugu Sakihara, "History of Okinawa," in *Uchinanchu: A History of Okinawans in Hawaii,* ed. Ethnic Studies Oral History Project, Ethnic Studies Program (Honolulu: University of Hawai'i at Manoa, 1981), 17–22.

9. The term "Ryukyuan" did not win public support; instead, "Okinawan" became commonly used.

10. Ruth Adaniya, "United Okinawan Association of Hawaii," in *Uchinanchu: A History of Okinawans in Hawaii,* ed. Ethnic Studies Oral History Project, Ethnic Studies Program (Honolulu: University of Hawai'i at Manoa, 1981), 330.

11. Hawai Nihonjin Iminshi Kankō Iinkai, ed., *Hawai Nihonjin iminshi* [A history of Japanese immigrants in Hawai'i] (Honolulu: United Japanese Society of Hawaii, 1962), 314.

12. Shimada Noriko, *Sensō to imin no shakaishi: Hawai nikkei Amerika-jin no Taiheiyō Sensō* [The social history of the war and the immigrants: The Pacific War of Japanese Americans in Hawai'i] (Tokyo: Gendai Shiryō Shuppan, 2004), 226–234.

13. Adaniya, "United Okinawan Association of Hawaii," 330–331; Okano Nobukatsu, "Sengo Hawai ni okeru 'Okinawa-mondai' no tenkai: Beikoku no Okinawa tōchi seisaku to Okinawa imin no kankei ni tsuite" [The "Okinawan problem" in Hawaii after World War II: U.S. occupation policy of Okinawa and Okinawan immigrants], *Imin Kenkyū* 4 (February 2008): 1–30, esp. 4–10.

14. Adaniya, "United Okinawan Association of Hawaii," 329.

15. Y. Baron Goto, "Trip Report to Okinawa, March 20, April 5; May 25–May 29, 1955," Records of the US Civil Administration of the Ryukyu Islands (USCAR), Okinawa Prefectural Archives, 1.

16. These four young fishermen were Seikō Agena, Buntarō Ishikawa, Takeharu Tōma, and Yukio Shin'ya.

17. From 1952 to 1970, 253 young farmers came to Hawai'i and received training. Sōmukyoku Shōgaika, "Okinawa ni enjo kyōryoku o shita Hawai kankei no omona shiryō" [The principal documents of relief from Hawai'i], December 8, 1970, Special Collection of the Okinawa Prefectural Archives.

18. Adaniya, "United Okinawan Association of Hawaii," 329.

19. Ibid.

20. Interview with Frank Goto, September 4, 2008. Later, Nagamine became the president of the *Ryūkyū Shinpō* and greatly supported the project.

21. Interview with Akira Ōtani, March 1, 2008.

22. Interview with Tokusaburō Uehara, September 25, 2008.

23. Interview with Hiroshi Nakashima, September 9, 2008.

24. Among them, Buntaro Ishikawa from Henza was invited to Hawai'i in 1953 as one of four young trainee fishermen. Therefore, it was his second visit.

25. "Yūbō na Hawai gyogyō seinen" [Promising young Okinawan fishermen], *Ryūkyū Shinpō*, January 23, 1962, 3.

26. Of these five men, three were from Henza, one was from Itoman, and one was from the island of Tsuken.

27. Naha, the capital city of Okinawa, had emerged as a fishing center after the war.

28. "Jūjitsu shikitta seikatsu" [Living a full life], *Ryūkyū Shinpō,* November 5, 1963, 7.

29. US Bureau of Commercial Fisheries, "Hawaii Fishery Training Program Prospectus" (Honolulu: Hawaii Area Office, US Bureau of Commercial Fisheries, n.d.), 28. The participants in the meeting were Albert J. Feirer, Hawaii Department of Education; William Fung, Hawaii State Employment Service; Herbert D. Hart, Hawaiian Tuna Packers; William Kanakanui, Hawaii Tuna Boat Owners Association; John C. Marr, US Bureau of Commercial Fisheries; Robert Robinson, Hawaii Department of Economic Development; Kenneth Sherman (chair), US Bureau of Commercial Fisheries; and Michio Takata, Hawaii Division of Fish and Game.

30. Candace L. Ball, "The Aku Fishermen of Honolulu: An Exploration of a Unique, Dying Lifestyle," Pacific Prize Contest Paper (1973), 2; Joseph R. Morgan, "Hawaii's Aku Fishery: Is It Dying?" (privately printed, 1974), 16.

31. Interview with Hisashi Itō, September 5, 2012.

32. Interview with Kiyoshi Tamashiro, September 23, 2008.

33. Interview with Tamotsu Ashitomi, August 29, 2009.

34. Henza Konjaku Shashinshū Henshū Iinkai, *Hiyamuza Kanamori,* 148–159.

35. Interview with Hirofumi Itō, September 30, 2009.

36. Interview with Tokusaburō Uehara, September 25, 2008.

37. Ōtani Matsujirō, *Waga hito to narishi ashiato: Hachijū nen no kaiko* [Becoming a man: Memoirs of my eighty years] (Honolulu: M. Ōtani, 1971), 130; P. V. Garrod and K. C. Chong, *The Fresh Fish Market in Hawaii* (Honolulu: Hawaii Agricultural Experiment Station, College of Tropical Agriculture, University of Hawai'i, 1978), 8–9.

38. Ichikawa, *Itoman gyogyō no tenkai kōzō*, 106–107.

39. Interview with Tokusaburō Uehara, September 25, 2008.

40. Interview with Kiyoshi Tamashiro, September 23, 2008.

41. Interview with Hirofumi Itō, September 30, 2009.

42. Interview with Yukikazu Onaga, September 23, 2008.

43. Young Cheng Shang, "The Skipjack Tuna Industry in Hawaii: Some Economic Aspects" (Honolulu: Economic Research Center, University of Hawai'i, July 1969), 9.

44. Richard N. Uchida and Ray F. Sumida, "Analysis of the Operation of Seven Hawaiian Skipjack Tuna Fishing Vessels, June–August 1967" (Seattle: National Marine Fisheries Service, 1971), 2.

45. Garrod and Chong, *The Fresh Fish Market in Hawaii*, 2, 16.

46. Interview with Tamotsu Ashitomi, August 29, 2009.

47. Interview with Seitoku Kinjō, August 29, 2009.

48. Ibid.

49. Interview with Hirofumi Itō, September 30, 2009.

50. Interview with Yukikazu Onaga, March 7, 2009.

51. At first, trainees received fixed wages regardless of their level of performance. However, this system soon changed as the result of many complaints from them. Interview with Shintoku Miyagi, June 16, 2010.

52. US Bureau of Commercial Fisheries, Hawaii Fisheries Training Program Prospectus, ii; Ball, "The Aku Fishermen of Honolulu," 7–13. According to Ball's observations, the earnings were divided in this way as of 1973: boat owner, 40 percent; captain, 7 percent of the owner's share (if he was not the owner); engineers, 2 percent of the owner's share; and the remaining 60 percent was divided equally among the crew members. The captain and the engineer shared equally in this portion, but the cook, usually a novice, earned 70 percent of a crew member share.

53. Interview with Hisashi Itō, September 5, 2012.

54. Interview with Hirofumi Itō, September 30, 2009.

55. Interview with Tokusaburō Uehara, September 25, 2008.

56. Interview with Hirofumi Itō, September 30, 2009.

57. Interview with Yasumasa Ōshiro, September 5, 2012.

58. Interview with Tamotsu Ashitomi, August 29, 2009.

59. Interview with Seitoku and Teru Kinjō, August 29, 2009; interview with Masaru Kinjō, June 17, 2010.

60. Interview with Ken Uehara, March 7, 2009; Interview with Kaoru Shingaki, March 7, 2009. For the general history and culture of Itoman, see Ueda Fujio, "Okinawa no tabi gyomin: Itoman gyomin no rekishi to seikatsu" [Migratory fishermen of Okinawa: The history and life of fishermen in Itoman]," in *Ryūkyūko no sekai* [The world of Ryukyu Arch], ed. Tanigawa Ken'ichi et al. (Tokyo: Shogakkan, 1992).

61. Interview with Seitoku and Teru Kinjō, August 29, 2009.

62. Interview with Masaru Kinjō, June 17, 2010.

63. Interview with Tamotsu and Masayo Ashitomi, August 30, 2009.
64. Interview with Hirofumi Itō, September 30, 2009.

Epilogue

1. Interview with Hiroshi Nakashima, September 9, 2008.
2. P. V. Garrod and K. C. Chong, *The Fresh Fish Market in Hawaii* (Honolulu: Hawaii Agricultural Experiment Station, College of Tropical Agriculture, University of Hawai'i, 1978), 2, 16; Division of Fish and Game, Department of Land and Natural Resources, State of Hawaii, *Executive Summary of the Hawaii Coastal Zone Fisheries Management Study* (Honolulu: Division of Fish and Game, Department of Land and Natural Resources, State of Hawaii, 1979), 26.
3. As of 1979, approximately 2,500 commercial fishermen were licensed with the State Division of Fish and Game. Most of them engaged in fishing on a part-time basis. Their median age was forty years, 84 percent lived in Hawai'i all of their lives, and non-citizens accounted for only 4 percent. Approximately 5 percent were female. Division of Fish and Game, Department of Land and Natural Resources, State of Hawaii, *Executive Summary of the Hawaii Coastal Zone Fisheries Management Study*, 6.
4. Chūgoku Shinbunsha, *Imin* [Immigration] (Hiroshima: Chūgoku Shinbunsha, 1992), 27.
5. In addition, handline fishing to catch bigeye and yellowfin tuna, in particular, prospered on the island of Hawai'i that had *Ahi kora,* or locations where yellowfin tuna are known to aggregate. Samuel G. Pooley, "Hawaii's Marine Fisheries" (Ethnic Studies Community Conference Papers, May 20, 1995), 1; Hawaii Seafood Project, *The Hawaii Fishing and Seafood Industry* (Honolulu: National Oceanic and Atmospheric Administration, 2007), 4; interview with Brooks Takenaka, September 4, 2008; interview with Frank Goto, September 4, 2008.
6. Thomas S. Hida and Robert A. Skillman, "A Note on the Commercial Fisheries in Hawaii," Southwest Fisheries Administrative Report (Honolulu: National Marine Fisheries Sercice, 1983), 1.
7. Hawaiian Tuna Packers, *Tuna: Hawaii's Harvest of the Sea* (Honolulu: Hawaiian Tuna Packers, n.d.).
8. Interview with Frank Goto, September 4, 2008.
9. Christofer H. Boggs and Bert S. Kikkawa, "The Development and Decline of Hawaii's Skipjack Tuna Fishery," *Marine Fisheries Review* 55, no. 2 (1993): 62.
10. Interview with Masami Shinzato, September 29, 2009.
11. Interview with Masami Shinzato, Tetsushi Tamashiro, Takeru Nakandakari, and Hiromasa Tamashiro, September 29, 2009.
12. Interview with Takeko Nakashima, October 3, 2009.
13. Interview with Hiroshi Nakashima, September 9, 2008, and August 17, 2009.
14. Interview with Takeko Nakashima, October 3, 2009.
15. Interview with Lisa Nakashima, October 3, 2009.
16. Ibid.
17. Ibid.
18. Interview with James N. Okuhara and Sueko Arakaki Okuhara, October 1, 2009.
19. Garrod and Chong, *The Fresh Fish Market in Hawaii,* 10–11, 14–15.

20. Interview with Takeko Nakashima, October 3, 2009.

21. "Hiro-shi no suisan (seri-ichi) heisa" [The close of Suisan fish auction at Hilo City], Hawaii123.com, http://www.hawaii123.com/news/01080102.html, accessed February 1, 2007.

22. James Gonser, "Fish Auction Set to Move," *Honolulu Advertiser,* July 6, 2004, http://the.honoluluadvertiser.com/article/2004/Jul/06/ln/ln10a.html, accessed April 29, 2014.

23. Hawaii Seafood Project, *The Hawaii Fishing and Seafood Industry,* 2.

24. Susan Blackmore Peterson, "Discussions in a Market: A Study of the Honolulu Fish Auction," PhD diss., University of Hawai'i, 1973, 116–119.

25. Brooks Takenaka and Leonard Torricer, *Trends in the Market for Mahimahi and Ono in Hawaii* (Honolulu: NOAA, 1984): 1–7.

26. Interview with Nobuo Tsuchiya, September 15, 2008.

27. Hawaii Seafood Project, *The Hawaii Fishing and Seafood Industry,* 3.

28. Interview with Hisao Shimizu and Shizue Shimizu, September 1, 2008.

29. The depletion of fish stocks due to overfishing is not peculiar to Hawaiian waters but is a serious concern in other major fishing grounds of the world, including Japan. Although this book does not explore ecological problems, other books discuss global trends in marine resources and fisheries: Mansel G. Blackford, *Pathways to the Present: U.S. Development and Its Consequences in the Pacific* (Honolulu: University of Hawai'i Press, 2007): Mansel G. Blackford, *Making Seafood Sustainable: American Experiences in Global Perspective* (Philadelphia: University of Pennsylvania Press, 2012).

30. Japan-America Society of Hawaii, Ehime Maru Memorial Association, http://www.jashawaii.org/emma.asp, accessed May 8, 2010.

Bibliography

Archives

Hawai'i State Archives, Honolulu
Okinawa Prefectural Archives, Haebaru town, Okinawa Prefecture
    Records of the US Civil Administration of the Ryukyu Islands (USCAR)
University of Hawai'i at Manoa, Honolulu
    Hawaii War Records Depository (HWRD)

Newspapers and Periodicals

Hawai Hōchi
Hawai'i Herald
Hawaii Times
Hawai Shokumin Shinbun
Honolulu Advertiser
Maui Shinbun
Nippu Jiji
Nippujiji Hawai Nenkan
Pacific Commercial Advertiser
Ryūkyū Shinpō
Yamato Shinbun

Interviews

Ashitomi, Masayo. Henza, Okinawa. August 30, 2009.
Ashitomi, Tamotsu. Henza, Okinawa. August 29, 2009; August 30, 2009.
Funai, Teruo. Honolulu. March 3, 2008.
Goto, Frank. Honolulu. September 4, 2008.
Itō, Hirofumi. Honolulu. September 30, 2009.
Itō, Hisashi. Honolulu. September 5, 2012.

Kida, Donald. Honolulu. September 9, 2008.
Kinjō, Fukuo. Itoman, Okinawa. September 7, 2007.
Kinjō, Masaru. Itoman, Okinawa. June 17, 2010.
Kinjō, Seitoku. Itoman, Okinawa. August 29, 2009.
Kinjō, Teru. Itoman, Okinawa. August 29, 2009.
Miyagi, Shintoku. Kyan, Okinawa. June 16, 2010.
Munaoka, Chizuko. Kanesaki, Fukuoka, Japan. January 28, 2009.
Munaoka, Tomio. Kanesaki, Fukuoka, Japan. January 28, 2009.
Nakandakari, Takeru. Honolulu. September 9, 2009.
Nakashima, Hiroshi. Honolulu. September 9, 2008; July 10, 2009.
Nakashima, Lisa. Honolulu. October 3, 2009.
Nakashima, Takeko. Honolulu. October 3, 2009.
Okuhara, James N. Honolulu. October 1, 2009.
Okuhara, Sueko Arakaki. Honolulu. October 1, 2009.
Onaga, Yukikazu. Itoman, Okinawa. September 25, 2008; March 7, 2009.
Ōshiro, Yasumasa. Honolulu. September 5, 2012.
Ōtani, Akira. Honolulu. September 4, 2007; March 1, 2008; March 3, 2008.
Ōtani, Evelyn. Honolulu. September 4, 2007.
Ōtani, Nancy. Honolulu. September 4, 2007.
Shimizu, Hisao. Honolulu. March 3, 2008.
Shimizu, Shizue. Honolulu. March 3, 2008; September 1, 2008.
Shingaki, Kaoru. Itoman, Okinawa. March 7, 2009.
Shinzato, Masami. Honolulu. September 29, 2009.
Tagawa, Hideo. Shimonoseki, Yamaguchi. December 14, 2009.
Takenaka, Brooks. Honolulu. September 4, 2008.
Tamashiro, Hiromasa. Honolulu. September 9, 2009.
Tamashiro, Kiyoshi. Itoman, Okinawa. September 23, 2008.
Tamashiro, Tetsushi. Honolulu. September 29, 2009.
Tsuchiya, Nobuo. Honolulu. September 15, 2008.
Uehara, Ken. Itoman, Okinawa. March 7, 2009.
Uehara, Tokusaburō. Itoman, Okinawa. September 25, 2008; March 7, 2009.

Secondary Sources

Adaniya, Ruth. "United Okinawan Association of Hawaii." In *Uchinanchu: A History of Okinawans in Hawaii*, edited by Ethnic Studies Oral History Project, Ethnic Studies Program, 324–336. Honolulu: University of Hawai'i at Manoa, 1981.

Ageno Kanzaburō. "Hawai no gyogyō o ronzu" [Talking about commercial fishing in Hawai'i]. In *Hawai Nihonjin jitsugyō shōkaishi* [The introduction of Japanese industries in Hawai'i], edited by Hawai Hōchi, 41–44. Honolulu: Hawai Hōchi, 1941.

———. "Yonjū-gonen mae no gyogyō-kai" [The fisheries industry forty years ago]. *Hawaii Times 60th Anniversary* no. 2 (October 1, 1955): 12.

Ai, Chung Kun. *My Seventy-Nine Years in Hawaii.* Hong Kong: Cosmorama Pictorial Publisher, 1960.

Azuma, Eiichiro. *Between Two Empires: Race, History, and Transnationalism in Japanese America.* Oxford: Oxford University Press, 2005.

Ball, Candace L. "The Aku Fishermen of Honolulu: An Exploration of a Unique, Dying Lifestyle." Pacific Prize Contest Paper, 1973.

Bell, Frank T., and Elmer Higgins. "A Plan for the Development of the Hawaiian Fisheries." Washington, DC: US Government Printing Office, 1939.

Bestor, Theodore C. *Tsukiji: The Fish Market at the Center of the World.* Berkeley: University of California Press, 2004.

Blackford, Mansel G. *Making Seafood Sustainable: American Experiences in Global Perspective.* Philadelphia: University of Pennsylvania Press, 2012.

———. *Pathways to the Present: U.S. Development and Its Consequences in the Pacific.* Honolulu: University of Hawai'i Press, 2007.

Boggs, Christofer H., and Bert S. Kikkawa. "The Development and Decline of Hawaii's Skipjack Tuna Fishery." *Marine Fisheries Review* 55, no. 2 (1993): 61–68.

Borgstrom, Georg. *Japan's World Success in Fishing.* London: Fishing News, 1964.

Brock, Vernon E. "A Proposed Program for Hawaiian Fisheries." *Hawaii Marine Laboratory Technical Report* no. 6 (1965).

Burton, Jeffery F., and Mary M. Farrell, *World War II Japanese Internment Sites in Hawai'i.* Honolulu: Japanese Cultural Center of Hawai'i, 2007.

Char, Tin-Yuke, ed. *The Sandalwood Mountains: Readings and Stories of the Early Chinese in Hawaii.* Honolulu: University of Hawai'i Press, 1975.

Chūgoku Shinbunsha. *Imin* [Immigration]. Hiroshima: Chūgoku Shinbunsha, 1992.

Clark, John R. K. *Guardian of the Sea: Jizo in Hawai'i.* Honolulu: University of Hawai'i Press, 2007.

Cobb, John N. "The Commercial Fisheries of the Hawaiian Islands in 1903." Department of Commerce and Labor, Bureau of Fisheries. Washington, DC: Government Printing Office, 1905.

Collins, Donald E. "Wirin, Abraham Lincoln." In *Encyclopedia of Japanese American History,* edited by Brian Niiya, 412–413. New York: Facts on File, 2001.

Division of Fish and Game, Department of Land and Natural Resources, State of Hawaii. *Executive Summary of the Hawaii Coastal Zone Fisheries Management Study.* Honolulu: Division of Fish and Game, Department of Land and Natural Resources, State of Hawaii, 1979.

Dole, S. B. "The Old Fish Market." *Twenty-Ninth Annual Report of the Hawaiian Historical Society* (1921): 19–25.

Erickson, Ethel. "Earnings and Hours in Hawaii Woman-Employing Industries." Bulletin of the Women's Bureau. Washington, DC: US Government Printing Office, 1940.

Ethnic Studies Oral History Project, Ethnic Studies Program, ed. *Uchinanchu: A History of Okinawans in Hawaii*. Honolulu: University of Hawai'i at Manoa, 1981.

Friday, Chris. *Organizing Asian American Labor: The Pacific Coast Canned-Salmon Industry, 1870–1942*. Philadelphia: Temple University Press, 1994.

Garrod, P. V., and K. C. Chong. *The Fresh Fish Market in Hawaii*. Honolulu: Hawaii Agricultural Experiment Station, College of Tropical Agriculture, University of Hawai'i, 1978.

Gibson, Arrell Morgan, with John S. Whitehead. *Yankees in Paradise: The Pacific Basin Frontier*. Albuquerque: University of New Mexico Press, 1993.

Gishi-wajinden, Gokansho-tōiden, Sōsho-wakokuden, Suisho-wakokuden. Tokyo: Iwanami Shoten, 1951.

Glazier, Edward, Janna Shackeroff, Courtney Carothers, Julia Stevens, and Russell Scalf. *A Report on Historic and Contemporary Patterns of Change in Hawai'i-Based Pelagic Handline Fishing Operations—Final Report*. Honolulu: School of Ocean and Earth Science and Technology, University of Hawai'i at Manoa, 2009.

Gotō Akira. "Hawai Nikkei imin no gyogō to Nanki-chihō no kenken gyohō" [The fishing gears of Japanese immigrants in Hawai'i and kenken fishing in the Nanki area]. *Mingu Kenkyū* 84 (1989): 1–6.

———. *Umi no bunka-shi* [The cultural history of the sea]. Tokyo: Miraisha, 1996.

Goto, Hisao, Kazuo Shinoto, and Alexander Spoehr. "Craft History and the Merging of Tool Traditions: Carpenters of Japanese Ancestry in Hawaii." *Hawaiian Journal of History* 17 (1983): 156–184.

Gulliver, Katrina. 2011. "Finding the Pacific World." *Journal of World History* 22, no. 1 (2011): 83–100.

Guthrie-Shimizu, Sayuri. "Occupation Policy and the Japanese Fisheries Management Regime, 1945–1952." In *Occupied Japan: The U.S. Occupation and Japanese Politics and Society*, edited by Mark E. Caprio and Yoneyuki Sugiki. New York: Routledge, 2007.

Haan, Aubrey, and Albert L. Tester. 1949. "Hawai'i's Fishing Industry." *Hawaii Educational Review* 38 (1949): 60–61, 70, 76.

Hamamoto, H. 1928. "The Fishing Industry of Hawaii." BA thesis, University of Hawai'i.

Hawaiian Tuna Packers. *Tuna: Hawaii's Harvest of the Sea*. Honolulu: Hawaiian Tuna Packers, n.d.

Hawai Nihonjin Iminshi Kankō Iinkai, ed. *Hawai Nihonjin iminshi* [A history of Japanese immigrants in Hawai'i]. Honolulu: United Japanese Society of Hawaii, 1962.

Hawai Shinpōsha. *Hawai Nihonjin nenkan* [Almanac of the Japanese in Hawai'i]. Honolulu: Hawai Shinpōsha, 1921.

Hawai Wakayama Kenjinkai, ed. *Fukkatsu jūgoshūnen kinenshi* [The fifteenth anniversary of the revival]. Honolulu: Hawai Wakayama Kenjinkai, 1963.

Hayase Shinzō. "Meiji-ki Manira-wan no Nihonjin gyomin" [Japanese fishermen in the Manila Bay during the Meiji period]. In *Kaijin no sekai* [The world of sea people], edited by Akimichi Tomoya, 343–368. Tokyo: Dōbunkan Shuppan, 1998.

Henza Konjaku Shashinshū Henshū Iinkai, ed. *Hiyamuza kanamori: Shashin ni miru Henza konjyaku* [The past and the present pictures of Henza]. Yonashiro: Henza Jichikai, 1984.

Hida, Thomas S., and Robert A. Skillman. *A Note on the Commercial Fisheries in Hawaii*. Southwest Fisheries Administrative Report. Honolulu: National Marine Fisheries Service, 1983.

Hiroshimaken, ed. *Hiroshimaken-shi minzoku-hen* [The history of Hiroshimaken, folklore]. Hiroshimashi: Hiroshimaken, 1978.

Hiroshimashi, ed. *Shinshū Hiroshimashi daisankan shakaikeizai-hen* [The new history of Hiroshima city, vol. 3. Society and economy]. Hiroshimashi: Hiroshima Shiyakusho, 1959.

Horne, Gerald. *The White Pacific: U.S. Imperialism and Black Slavery in the South Seas after the Civil War*. Honolulu: University of Hawai'i Press, 2007.

Ichikawa Hideo. *Itoman gyogyō no tenkai kōzō* [The development and structure of fishing in Itoman]. Naha: Okinawa Times, 2009.

Iida Kōjirō. *Hawai Nikkeijin no rekishi chiri* [The history and geography of Japanese Americans in Hawai'i]. Kyoto: Nakanishiya Shuppan, 2003.

Ishigaki Ayako. *Ai to Wakare* [Love and farewell]. Tokyo: Kōbunsha, 1958.

Ishikawa, Tomonori. "Historical Geography of Early Immigrants." In *Uchinanchu: A History of Okinawans in Hawaii*, edited by Ethnic Studies Oral History Project, Ethnic Studies Program, 80–104. Honolulu: University of Hawai'i at Manoa, 1981.

———. *Nihon imin no chirigakuteki kenkyū* [The geographical study of Japanese immigrants]. Ginowan: Yōju Shoin, 1997.

Itomanshi-shi Hensan Iinkai, ed. *Itomanshi-shi: Shiryō-hen 7, senjishiryō jōkan* [History of Itoman: Historical document 7, wartime records, vol. 1]. Itoman: Itoman Shiyakusho, 2003.

Iwashita, T. T. "Development and Design of the Hawaiian Fishing Sampans." Paper presented at the Hawaii Section of the Society of Naval Architects and Marine Engineers, 1956.

Jenkins, Oliver P. *Report on Collection of Fishes Made in the Hawaiian Island, with Description of New Species*. Washington, DC: Government Printing Office, 1903.

Jones, Ryan Tucker. "Running into Whales: The History of the North Pacific from below the Waves." *American Historical Review* 118 (2013): 349–377.

Kahā'ulelio, Daniel. *Ka'Oihana Lawai'a': Hawaiian Fishing Traditions*. Honolulu: Bishop Museum Press, 2006.

Kanezaki Gyogyō-shi Hensan Iinkai, ed. *Chikuzen Kanezaki gyogyō-shi* [The history of Kanezaki Fishery in Chikuzen]. Fukuoka: Kanezaki Gyogyō Kyōdō Kumiai, 1992.

Kawakami Masayuki. *Hiroshima Ōtagawa deruta no gyogyōshi* [The fisheries history of the Hiroshima Ōta River delta]. Hiroshimashi: Takumi Shuppan, 1976.

Kawamoto, Kurt E., Russell Y. Ito, Raymond P. Clarke, and Allison A. Chun. "Status of the Tuna Longline Fishery in Hawaii, 1987–88." Southwest Fisheries Center Administrative Report. Honolulu: National Marine Fisheries Service, 1989.

Kawaoka Takeharu. *Umi-no-tami: Rekishi to minzoku* [People of the sea: Their history and customs]. Tokyo: Heibonsha, 1987.

Kida Katsukichi. "Waga chichi o kataru" [Talking about my father]. In *Fukkatsu jūgoshūnen kinenshi* [The fifteenth anniversary of the revival], edited by Hawai Wakayama Kenjinkai, 198–199. Honolulu: Hawai Wakayama Kenjinkai, 1963.

Kim Byungchul. *Ebune no minzokushi: Gendai Nihon ni ikiru umi no tami* [The history of sea nomads: The sea people living in contemporary Japan]. Tokyo: Tokyo Daigaku Shuppan, 2003.

———. "Sea Nomads of Japan." *International Journal of Maritime History* 11, no. 2 (1999): 87–105.

Kodama-Nishimoto, Michi, Warren S. Nishimoto, and Cynthia A. Oshiro, eds. *Hanahana: An Oral History Anthology of Hawaii's Working People*. Ethnic Studies Oral History Project. Honolulu: University of Hawai'i at Manoa, 1984.

Konishi, Owen K. "Fishing Industry of Hawaii with Special Reference to Labor." Honolulu: University of Hawai'i Reports of Students in Economics and Business, 1930.

Kushimotochō-shi Hensan Iinkai, ed. *Kushimotochō-shi* [The history of Kushimotochō]. Kushimotochō: Kushimotochō Yakuba, 1995.

Lydon, Sandy. *The Japanese in the Monterey Bay Region: A Brief History*. Capitola, CA: Capitola, 1997.

Maeda Takakazu. *Hawai no jinjya-shi* [The history of Shinto shrines in Hawai'i]. Tokyo: Taimeidō, 1999.

Manu, Moke, et al. *Hawaiian Fishing Traditions*. Honolulu: Kalamakū Press, 2006.

Markrich, Mike. "Fishing for Life." In *Kanyaku Imin: A Hundred Years of Japanese Life in Hawaii*, edited by Leonard Lueras, 142–143. Honolulu: International Savings and Loan Association, 1985.

Marlatt, Daphne. *Steventon Recollected: Japanese-Canadian History*. Vancouver: Provincial Archives of British Columbia, 1975.

Matsumoto Sei. "Hawai no shokuryō mondai to suisan gakkō no setsuritsu" [The food problem in Hawai'i and establishment of a fisheries school]. In *Hawai Nihonjin jitsugyō shōkaishi* [The introduction of Japanese industries in Hawai'i], edited by Hawai Hōchi, 47–48. Honolulu: Hawai Hōchi, 1941.

Miwa Chitoshi. "Hiroshimaken-ka Setonaikai mikaihou buraku gyofu no senkai shutsuryō" [Going out fishing in Korean waters of buraku fishermen in Hiroshima Prefecture of Setonaikai]. *Gyogyō Keizai Kenkyū* 21, no. 2 (1975): 43–51.

Miyamoto Tsuneichi. *Setonaikai no kenkyū* [The study of Seto Inland Sea]. Tokyo: Miraisha, 1965.

———. *Tsushima gyogyō-shi* [The history of fisheries in Tsushima]. Tokyo: Miraisha, 1983.

Morgan, Joseph R. "Hawaii's Aku Fishery: Is It Dying?" Privately printed, 1974.

Morimoto Takashi. *Tōwachō-shi: Kakuron-hen dai sankan gyogyōshi* [History of Tōwachō. Vol. 3, The record of fisheries]. Suō-Ōshimachō: Tōwachō Yakuba, 1986.

Morita Sakae. *Hawai Nihonjin hatten-shi* [The development of Japanese history in Hawaiʻi]. Waipahu: Shin'eikan, 1915.

Nadel-Klein, Jane, and Dona Lee Davis. *To Work and to Weep: Women in Fishing Economies*. St. John's: Institute of Social and Economic Research, Memorial University of Newfoundland, 1988.

Nakadate Kou, ed. *Nihon ni okeru kaiyōmin no sōgōteki kenkyū, jōkan* [The general study of the sea people in Japan, vol. 1]. Fukuoka: Kyushu Daigaku Shuppankai, 1987.

———, ed. *Nihon ni okeru kaiyōmin no sōgōteki kenkyū, gekan* [The general study of the sea people in Japan, vol. 2]. Fukuoka: Kyushu Daigaku Shuppankai, 1989.

Nippu Jijisha. *Kanyaku imin Hawai tokō gojusshūnen kinenshi* [The special volume commemorating the fiftieth anniversary of the arrival of the government contract immigrants]. Honolulu: Nippu Jijisha, 1935.

Nōshōmushō Suisankyoku, ed. *Kaigai ni okeru honpō hōjin no gyogyō jōkyō* [The overseas fishing of the Japanese]. Tokyo: Nōshōmushō Suisankyoku, 1918.

———. *Nihon suisan hosaishi* [The record of Japanese fishing]. 1911–1935. Reprint, Tokyo: Iwasaki Bijutsusha, 1983.

Odo, Franklin. *No Sword to Bury: Japanese Americans in Hawaiʻi during World War II*. Philadelphia: Temple University Press, 2004.

Ogawa, Dennis M. *Kodomo No Tame Ni: For the Sake of the Children*. Honolulu: University of Hawaiʻi Press, 1978.

Ogawa, Dennis M., and Evarts C. Fox Jr. "Japanese Internment and Relocation: The Hawaii Experience." In *Japanese Americans: From Relocation to Redress*, edited by Roger Daniels, Sandra Taylor, and Harry H. L. Kitano, 135–138. Seattle: University of Washington Press, 1986.

Ogawa Taira. *Arafurakai no shinju* [Pearls of Arafura Sea]. Tokyo: Ayumi Shuppan, 1976.

Okano Nobukatsu. "Sengo Hawai ni okeru 'Okinawa-mondai' no tenkai: Beikoku no Okinawa tōchi to Okinawa imin mondai no kankei ni tsuite" [The "Okinawan problem" in Hawaii after World War II: U.S. occupation policy of Okinawa and Okinawan immigrants]. *Imin Kenkyū* 4 (2008): 1–30.

Okihiro, Gary Y. *Cane Fires: The Anti-Japanese Movement in Hawaiʻi, 1865–1945*. Philadelphia: Temple University Press, 1991.

Okihiro, Michael M., and Friends of Aʻala. *Aʻala: The Story of a Japanese Community in Hawaii*. Honolulu: Japanese Cultural Center of Hawaiʻi, 2003.

Ōkubo Gen'ichi, ed. *Hawai Nihonjin hatten meikan bōchōban* [The biographical record of the development of the Japanese in Hawai'i, Bōchō edition]. Honolulu: Hawai Shōgyōsha, 1940.

Okumura Takie. *Hawai ni okeru nichibei mondai kaiketsu undō* [The movement for the solution of Japan-US problems in Hawai'i]. Kyoto: Naigai Shuppan Insatsu, 1925.

Onodera Tokuji, ed. *Honolulu Nihonjin Shōgyō Kaigisho nenpō* [The annual report of the Japanese Chamber of Commerce in Honolulu]. Honolulu: Honolulu Nihonjin Shōgyō Kaigisho, 1922.

Ōtani Matsujirō. "Nikkei gyogyō kaisha no hensen o kataru" [Talking about the development of the Japanese fishing companies]. *Hawaii Times 60th Anniversary* no. 9 (October 1, 1955): 10–11.

———. *Waga hito to narishi ashiato: Hachijū nen no kaiko* [Becoming a man: Memoirs of my eighty years]. Honolulu: M. Ōtani, 1971.

Peattie, Mark R. *Nan'yō: The Rise and Fall of the Japanese in Micronesia, 1885–1945*. Honolulu: University of Hawai'i Press, 1988.

Peterson, Susan Blackmore. "Decisions in a Market: A Study of the Honolulu Fish Auction." PhD diss., University of Hawai'i, 1973.

Saiki, Patsy Sumie. *Ganbare! An Example of Japanese Spirit*. Honolulu: Kisaku, 1982.

Sakihara, Mitsugu. "History of Okinawa." In *Uchinanchu: A History of Okinawans in Hawaii*, edited by Ethnic Studies Oral History Project, Ethnic Studies Program, 3–22. Honolulu: University of Hawai'i at Manoa, 1981.

Schug, Donald M. "Hawai'i's Commercial Fishing Industry: 1820–1945." *Journal of Hawaiian History* 35 (2001): 15–34.

Segawa Kiyoko. *Ama* [Women divers]. Tokyo: Miraisha, 1970.

———. *Hisagime* [Female peddlers]. Tokyo: Miraisha, 1971.

Seki Reiko. "Kaihatsu no umi ni shūsan suru hitobito: Henza ni okeru gyogyō no isou to mainā-sabushisutensu no tenkai" [The people who meet and part at the sea and development: The phase of fisheries on Henza and the development of minor subsistence]. In *Okinawa Rettō: Shima no Shizen to Dentō no Yukue* [The Okinawan archipelago: Future of the islands' nature and tradition], edited by Matsui Ken, 127–166. Tokyo: Tokyo Daigaku Shuppan, 2004.

Shang, Young Cheng. "The Skipjack Tuna Industry in Hawaii: Some Economic Aspects." Honolulu: Economic Research Center, University of Hawai'i, 1969.

Shimada Noriko. *Sensō to imin no shakaishi: Hawai nikkei Amerika-jin no Taiheiyō Sensō* [The social history of the war and the immigrants: The Pacific War of Japanese Americans in Hawai'i]. Tokyo: Gendai Shiryō Shuppan, 2004.

Shimizu Akira, ed. *Kinan no hitobito no kaigai taiken kiroku 1* [The recorded experiences of Kinan people in a foreign land, vol. 1]. Privately printed, 1993.

———, ed. *Kinan no hitobito no kaigai taiken kiroku 2* [The recorded experiences of Kinan people in a foreign land, vol. 2]. Privately printed, 1993.

———, ed. *Kinan no hitobito no kaigai taiken kiroku 3* [The recorded experiences of Kinan people in a foreign land, vol. 3]. Privately printed, 1993.

Sissons, David C. S. "1871–1946 nen no Ōsutoraria no Nihonjin" [The Japanese in Australia between 1871 and 1946]. *Ijūkenkyū* 10 (1974): 27–54.

Smith, Andrew F. *American Tuna: The Rise and Fall of an Improbable Food.* Berkeley: University of California Press, 2012.

Sōga Yasutarō. *Gojūnenkan no Hawai kaiko* [Memoirs of fifty years in Hawaiʻi]. Honolulu: Gojūnenkan no Hawai Kaiko Kankōkai, 1953.

Susamichōshi Hensan Iinkai, ed. *Susamichōshi 2* [The history of Susamichō, no. 2]. Susamichō, Wakayama: Susamichō, 1978.

Tajima Yoshiya. "Kita no umi ni mukatta Kishū shōnin" [The Kishū merchants who went to the northern lands]. In *Nihonkai to hokkoku bunka* [The Sea of Japan and northern culture], edited by Amino Yoshihiko. Tokyo: Shogakkan, 1990.

Takeda Naoko. *Manira e watatta setouchi gyomin* [Fishermen of Seto Inland Sea who went to Manila]. Tokyo: Ochanomizu Shobō, 2002.

Takenaka, Brooks, and Leonard Torricer. *Trends in the Market for Mahimahi and Ono in Hawaii.* Honolulu: NOAA, 1984.

Tasaka, Jack Y. "Hawai to Wakayama kenjin" [Hawaiʻi and people from Wakayama Prefecture]. *Journal of the Pacific Society* 31 (1986): 52–72.

Tilburg, Hans Konrad Van. "Vessels of Exchange: The Global Shipwright in the Pacific." In *Seascapes: Maritime Histories, Littoral Cultures, and Transoceanic Exchanges,* edited by Jerry H. Bentley, Renate Bridenthal, and Kären Wigen, 38–52. Honolulu: University of Hawaiʻi Press, 2007.

Titcomb, Margaret. *Native Use of Fish in Hawaii.* Honolulu: University of Hawaiʻi Press, 1972.

Uchida, Richard N., and Ray F. Sumida. "Analysis of the Operation of Seven Hawaiian Skipjack Tuna Fishing Vessels, June–August 1967." Seattle: National Marine Fisheries Service, 1971.

US Bureau of Commercial Fisheries. *Hawaii Fishery Training Prospectus.* Honolulu: Hawaii Area Office, US Bureau of Commercial Fisheries, n.d.

Ushiban Seikichi. "Hawai imin [Immigration to Hawaiʻi]." In *Furuki o tazunete* [Cherishing of the old], edited by Henza Jichikai, 345–348. Henza: Henza Jichikai, 1985.

Wakayamaken, ed. *Wakayamaken iminshi* [The immigration history of Wakayama Prefecture]. Wakayamashi: Wakayamaken, 1957.

Wakukawa, Earnest. *A History of the Japanese People in Hawaii.* Honolulu: Tōyō Shoin, 1938.

Yanagisawa Ikumi. "'Shashin hanayome' wa 'otto no dorei' dattanoka: 'shasin hanayome' tachi no katari o chūshin ni" [Was a 'picture bride' a slave of her husband? Narratives of picture brides]. In *Shashin hanayome, Sensō hanayome no tadotta michi: Josei-iminshi no hakkutsu* [Crossing the ocean: A new look at the history of Japanese picture brides and war brides], ed. Shimada Noriko, 47–85. Tokyo: Akashi Shoten, 2009.

Yokemoto Masafumi. "Senzen-ki Taiwan ni okeru Nihonjin gyogyō imin" [Japanese fishing immigrants in Taiwan before World War II]. *Tokyo Keidai Kaishi* 245 (2005): 95–111.

Yoneyama Hiroshi. "Amerika-shi jojyutsu no ekkyō-ka to Nihonjin no kokusai idō: Iminshi no wakugumi no kaitai to saikōchiku ni mukete" [The transnationalization of the description of American history and the transnational movements of the Japanese: Toward the destruction and reconstruction of the contours of immigration history]. *Ritsumeikan Bungaku* 597 (2007): 144–153.

———. "Kan-Taiheiyō chiiki ni okeru Nihonjin no idousei o saihakken suru" [Rediscovering the transpacific mobility of the Japanese]. In *Nikkeijin no keiken to kokusai idō: Zaigai Nihonjin imin no kingendaishi* [Transnational Japanese mobility in the modern era: The experiences of overseas Japanese and their descendants], edited by Yoneyama Hiroshi and Kawahara Norifumi, 9–23. Tokyo: Jinbun Shoin, 2007.

Yoshida Keiichi. *Chōsen suisan kaihatsu-shi* [On the development of the Korean fishery]. Shimonoseki: Chōsuikai, 1954.

Index

Note: Page numbers in *italics* indicate figures and maps.

Aʻala Market, 50, 81, 102, *103,* 117, 139, *139,* 158
Achu (Chinese merchant), 51–52
Ageno, Kanzaburō, 37, 87–88
agyā net fishing, 130–131, 183n1
ʻ*ahi* tuna, 38–39, 66, 70–71, 112–113
Ai, Chung Kun, 44
Ainu people, 12, 14
Aki (Hiroshima Prefecture), 15, *16,* 18–19, 21
aku tuna. *See* skipjack (*aku*) tuna
Alaska, 96
Alexander and Baldwin, 6
ama (women divers), 25–26
Amaterasu (sun goddess), 25
American Factors, 6, 58–59, 64
Asano, Nagaakira, 15, 18
Asari, Kuniyoshi "James," 101, 117
Asari, Walter, 65
Asato, Sadao, 135
Ashitomi, Masayo, 150–151
Ashitomi, Tamotsu, 140–141, 146, 149, 150–151
auction system, in Hawaiʻi, 50–52, 160–162
Australia, Wakayama fishermen and, 4, 28–31, 38

awa (milkfish), farming of, 90
Azumi-no-muraji (fishermen's clan), 11
"Back to the Land" slogan, 87, 89
"Back to the Sea" movement, 88–89, 99–100
Bagley, D. W., 113
bait fishing, 67–70, 95–96, *145.* See also ʻ*iao; nehu*
Banshū-maru (ship), 127–128, 163, 183n26
Beyer, Paul, 95, 97, 98
Black (Kuroshio) Current, 2, 11
Bōsō Peninsula, 12
Brewer and Co., 6
Broadbent, F. W., 110
Brooke, George, 92
Bunka era (1804–1817), 19
buraku residents, 20–21

California, 55–56, 95, 174n82
Canada, salmon fishing in, 4, 31–32, 76
canning: of salmon, 73, 107; of skipjack tuna, 63–64; of tuna in California, 55–56, 174n82
Cary, Miles, 88
Castle and Cooke, 6
Cayetano, Ben, 161
child care concerns, 9, 78, 80

199

China: competition and cooperation with Japanese, 43–45, 47–52; Japan's military aggression and, 96; net fishing in early twentieth century, 34, 36; Okinawa fishing and, 129–130, 132
Chinatown Fire (1900), 33, 45
Chōsen-gumi fleet, 21–22, 41
Chōshū (now Yamaguchi Prefecture), 17–18, *17*
citizenship issues, post–World War II, 124–126
City Market, 44
City of Tokio (ship), 34
clams, farming of, 90–91
Clark, Tom C., 123
Coelho, William J., 55, 56
crabs, transplantation of, 91

Davis, Dona Lee, 26
Davis, James J., 92
Dillingham, Walter F., 109–111, 112
Dillingham Corporation, 81
Distant Water Fishery Promotion Act (1897), 22
Dole, Sanford B., 50

Ebisu (god), 84
Ebisu Shrine, 84, *85*
ebune (house ship), 24–26
edamura (branch villages), 5, 28
Egawa, Heitarō, 43
Ehime Maru accident and memorial, 164–165, *164*
Ehime Prefecture, 27, 165
Environmental Protection Agency, 155
espionage, suspicions about Japanese fleet and, 98–99, 101, 109
Ezo (now Hokkaido), 4, 12, *13,* 14–15
Federal Surplus Commodities Corporation (F. S. C. C.), 112, 182n66
Fish Dealers' Guild, 51
Fishermen's Association, in Kaka'ako, 51, 83–84
fishing culture, development of Japanese, 11–32; characteristics of fishing communities, 27–28; before modern era, 11–19; opening of Japan and expansion of, 19–24; women's fishing and trading, 24–28
fishing industry, in Hawai'i. *See* Hawai'i
Fleming, David, 110, 111–113
Fong, Hiram L., 117
frozen food operations, begun after World War II, 118–119
Fukuoka Prefecture, 24–26, 29
Funadama (goddess), 24
Funai, Kimi, 66, 78
Funai, Seiichi, 66, *67,* 71, 124, 155
Funai, Teruo, 71, 78

Geishin-ya Ryokan inn, 45
gender. *See* women
Gotō, Akira, 29
Goto, Frank, 136, 137
Goto, Y. Baron, 135
Gotō Archipelago, 19, 21, 26
Gulf Oil Co., 141

Hachiman (god), 85
Hakuseiji Temple, 17, 84
Hama, Naoichi, 30–31
Hamaguchi, Banji, 40
Hamamoto, Yoshio, 89
haole (white) plantation owners, 52, 60, 87, 89, 94, 108, 110
Hara, Kanjirō, 21
Harding, Warren, 92
Hawai'i: anti-Japanese regulations and legislation, 54–57; contemporary fish consumption, 162; contemporary status of industry and fish stocks, 163–165; establishment of Japanese fishing companies and support of fishermen, 45, 47–51; factionalism among fishermen, 51–54; industry after Japanese domination, 153–165, 187n3; Japanese domination of operations, 60–61; Japanese family life and, 71–72, 74–75; Japanese fishing fleet, 86, 178n67; Japanese fishing styles and methods, 66–71; Japanese percentage of fish catch, in 1970s and 1980s,

153–154; licensing of Japanese fishermen and boats, 6, 52, 97, 101, 110, 113–114, 187n3; overfishing, depletion of stock, and new aquaculture, 89–91; power boats and expansion of fisheries, 62–64, 86, 178n67
Hawaiian Labor Commission, 92
Hawaiian Tuna Packers, Ltd., *77;* closing of, 155; Okinawan fishing training and, 139–140; origins of, 64; pre–World War II regulation of, 95–96, 98–99; women employed by, 75, *76, 77*–78; after World War II, 117; during World War II, 106–109, 112
Hawai'i Community Development Authority, 160–161
Hawaii Fishery Training Committee, 139
Hawaii Fishing Co., 45, 47, 50, 53, 60; boycott of, 51–52
Hawaii Island Fishing Co., 57, 178n67
Hawaii-Okinawa Farm Youth Training Program, 135
Hawaii Suisan Co., 53, 59–60
Hawaii Tuna Boat Owners Association, 138–140
Hawai Shokumin Shinbun, 57
Hayashi, Torazuchi, 90
Hayashikane Shōten, 120
Heen, William H., 117, 128
Henza (Okinawa), 131–132, 135, 138, 140–141, 143–144, 146–147, 150–151
Higher Wage Association, 45
Hills, Alfred D., 110
Hilo, 42–43, *44,* 60, 65, 82, 86, 122; boat repair in, 63; commercial fishing population in, 154; community life in, 72; contemporary fishing industry, 160; net fishing prohibited in Hilo Bay, 56–57; tsunami in, 118
Hilo Daijingu shrine, 122, 123, 124
Hilo (Waiakea) Suisan Co., 43, 56–57, 58, 64, 90, 93, 114, 118, 160, 175n8, 178n67
Hind, Leighton, 110
Hirabayashi, Gordon, 125

Hiroshima Prefecture: early fishing, 15, *16,* 18–19, 21; longline fishermen from, 70–71; travel to Australia from, 29; travel to Hawai'i from, 31–32, 34, 41–51; travel to Korean peninsula from, 21–23, 25, 169n16; travel to Philippines from, 23–24
Hirota, Hitoshi, 84
Honolulu, *46, 54,* 178n67
Honolulu Fish Auction, of United Fishing Agency, 154, 160–162, *161*
Honolulu Fishing Co., 47–48, 50, 53–54, 97, 178n67
Honolulu Tuna Fishermen's Association, 119
Honshu Island, *13*
Houston, Victor S. K., 94

'iao (silverside), 33, 36, 57, 67
ice, importance of, 62, 63, 70, 71, 80
Ichimatsu, Kumao, 40
Ichuman-ui system, 183n1
ie (patriarchal household system), 7–8
Ige, Chōichi, 135
ika (squid)-*shibi* (yellowfin tuna) fishing, 184n5
Iki (island), 19, 21, 26, 28
ina (rock-boring sea urchin), 90
internment, of Japanese fishermen during World War II, 101–102, 114, 115, 116
Ishikawa, Dr. Chiyomatu, 90
Islander (boat), 111–112
Islander Fishing Company, 112
issei (first-generation Japanese Americans), 3, 10, 88–89, 119, 163; peasant status of, 5–6; *Shunkotsu-maru* and, 120–123
Itō, Hirofumi, 141, 143–144, 147–148, 151–152
Itō, Hisashi, 140, 147
Itoman (Okinawa), 129–132, *137,* 138, 140, 142, 146–150, 158

Japan: in 2010, *13;* geographic conditions of, 1–2; maritime culture in, 1, 3–8; post–World War II rebuilding of ties with, 120–124; transnational migratory flows, 3–5

Japanese Americans. *See* issei; nisei
Jenkins, Oliver P., 90

Kadota, Kikumatsu, 43
kaifu (fishermen), 11
Kakaʻako, 45, 53, 58, 114, 154–155; boat building in, 63–64; community life in, 71–72; contemporary redevelopment of, 160–161; women in, 77–78
Kakaʻako Fishermen's Association, 51
Kalland, Arne, 169n10
kamaboko production, 63, 150, 158
Kamaishi, 12
Kamuro (journal), 86
Kanai, Zenzō, 154
Kāneʻohe Bay, 95, 99, 144
Kanezaki (Fukuoka Prefecture), 24–26
Kannon (goddess), 84–85
Kanto area, 12, 15
kanyaku imin (government-contract immigrants), 4
kapu (taboos), 36
Kashiwabara, Kiyoshi, 63
Kasuga-maru (boat), 57
Kasuga-maru I and *Kasuga-maru II*, 48
katagi (women fish peddlers), 7, 26–27, 34, 80, *81, 137,* 150
katsuobushi production, 63
Kauaʻi, 34, *74;* fishermen and fishing communities, 42, 72, 131; fishing fleet, 178n67; percentage of fish catch, in 1970s and 1980s, 154; stocking of, 90–91
Kelly, H. L., 95
kenken hook, 38
Kewalo Basin, 45, 48, 50, 53, 71, *72,* 102, 109, 114, 154, 155
Kida, Donald, 102, 116
Kida, Katsukichi, 65–66, 97–98, 102, 104, 114, 115, 116–117
Kida, Kiichi, 100
Kida, Matsu, 100
Kida, Sutematsu, 100
Kida, Tsurumatsu, 48, 57, 64, 65–66, 97–98
Kida, Yasue, 102, 116–117
Kīlauea Military Camp, 114

Kinan-maru (ship), *67,* 124, *125,* 126, 163
King Fishing Co., Ltd., 117–118
King Fish Market, 47, 58
Kinjō, Fukuo, 155
Kinjō, Masaru, 150
Kinjō, Seitoku, 146–147, 149
Kinjō, Teru, 149–150
Kishū (Wakayama Prefecture), 11–12, 14–16, *14,* 18, 19
Kishū-katsuosengumi (Kishū Skipjack Tuna Fishing Fleet), 41
Kitagawa, Isojirō, 42, 43, 58, *85*
Kitamae ship lines, 4, 15
kō (credit associations), 43, 58, 84–85
Komine, Heikichi, 40–41, 49, 113
konohiki rights, 174n79
Korean Peninsula: crowded fishing conditions in early twentieth century, 41–42; early Japanese fishermen's travel to, 18, 19, 25, 28; fishing methods in, 169n16; Japan's annexation of, 22–23; Meiji era travel to, 21–23, 169n16
Korematsu, Fred, 125
Korol, Alex, 88, 109–110
Kotohira-gū (shrine), 84
Kotohira Shrines, 83–84, 122–124
Kula Kai (boat), 155
kumiai cooperative, 49
Kuno, Gihei, 31–32
Kurihara, Nobu, 34
Kuril Islands, 12, 27, 28
Kuroshio (Black) Current, 2, 11
Kyodo Gyogyo (fishing cooperative), 117–118
Kyushu, 4, *13,* 15, 18–19, 22, 24, 41, 129

Lānaʻi, 39, 72, 154, 178n67
licensing, of Japanese fishermen and boats, 6, 52, 97, 101, 110, 113–114, 187n3
Lisa I (boat), 153, 157
lobsters, 154–155, 187n5
longline fishing, 19, 22, 23, 38–39, 66, 68–71, 124, 138, 154–155

MacArthur, Douglas, 120
Macfarlane, Walter J., 63–64, 99–100
mahimahi (dolphinfish), 70, 154, 162
Makino, Kinzaburō, 57
malau (bait holder), 36
Mamiya, Rinzō, 14
Masagatani, Yozō, 100, 102
Matsujirō Ōtani, Ltd., 114, 117; women and, 81–82, *82*
Matsukata, Masayoshi, 20
Matsuno, Kamezō, 42–43, 57, 114, 118
Matsuno, Rex, 114, 118
Maui, 37, 39, 42, 57, 72, *75*, 90, 111–113, 154, 178n67
Meiji era (1868–1912), 4, 8; modernization, militarization, and loss of fishing grounds during, 20–24
Mexico, 95
Miho, Katsurō, 125
Mitamura, Toshiyuki, 45, 47
Mitchell, Glen, 110
Miyagi, Shintoku, *139*
Molokaʻi, 41, 87, 154, 178n67
Murakami, Mankichi, 32
Mutsu, Munemitsu, 30

Nadel-Klein, Jane, 26
Nagamine, Hikomasa, 136, 185n20
Nagasaki Prefecture, 29, 169nn14, 16
Nakabe, Kenʻichi, 120, *121*
Nakabe, Tōjiro, 120
Nakafuji, Chōzaemon, 47–48
Nakamine, Shinsuke, 134, 137
Nakamura, Umatarō, 34
Nakashima, Allen, 157–158
Nakashima, Hiroshi, 138, *143*, 153, 156–158
Nakashima, Laurally, *159*
Nakashima, Lisa, 157–158, *159*
Nakashima, Takeko, 156–160, *159*
Nakasone, Fujiko, 148–149
Nakasuji, Gorokichi, 31–33, 36–39, 48, 53–54, 62, 70
Nakayama, Tatsunosuke, 21
Nan'yōmaru (ship), 32
National Fisheries University, 120, 122–123, *122*

National Marine Fisheries Service (NMFS), 144
Native Hawaiians: aquatic knowledge shared with Japanese, 38; decline in fishing by, 39; history of, 36; Japanese conflicts with methods and culture of, 36–37; self-sufficiency of, 37
nehu (Hawaiian anchovy), 33, 57, 67, 112, 144–145, *145*
net fishing, 23, 34–36, 38, 56, 66, 68–69, 108, 130–131, 183n1
Nippu Jiji (newspaper), 88, 96, 99
Nisei (boat), 155–156
nisei (second-generation Japanese Americans), 6, 105–106, 119, 123–126; changed attitudes toward fishing, 9–10, 86–89
Nonami, Kojirō, 29

Oʻahu, 39, 45, 57, 64, *73*, 90–91, 108, 153–154
Oʻahu Fish Market, 44–45
Oʻahu Market, 50, 156–157
Oahu Railway and Land Co., Ltd., 81
Odo, Franklin, 3
Office of Food Control, 106–107
Office of Food Production, 106–110, 113–114
ʻohana (extended family of communities), 37
Ōhara, Jinkurō and wife, 34
Okhotsk (Kurile) Current, 2
Oki (island), 26
Okikamuro (Yamaguchi Prefecture): development of pole-and-line fishing in, 16–18; fishermen from, 19, 23, 41–43, 53, 84–85; Meiji modernization and, 21–22
Okinawa, 4, 129–152, *130;* cultural differences with Hawaiʻi, 146–149; experiences of fishermen in Hawaiʻi, 142–149, 186nn51–52; separation of families and, 148–152; tensions with Japanese, 133–134; training programs for young fishermen, 134–141
Okino-Torishima island, 90
Okuhara, James N., 158
Okuhara, Sueko, 158

Okumura, Takie, 86–87, 89
omiyage gifts, 63, 175n6
Onaga, Yukikazu, 144, 147
opelu (mackerel), 68–69, 70, 80
Oregon, 55
Organic Act, 39
Oshika Peninsula, 12
Ōtani, Akira, 71, 79–80, 81, 136, 138
Ōtani, Florence, 81, *82*
Ōtani, Gladys, 81, *82*
Ōtani, Jiroichi, 81, *82*
Ōtani, Kane, 78, *103,* 104
Ōtani, Matsujirō, *59,* 78, *82, 103,* 124, *125,* 131, 136, 138; background, 58; challenge to Big Five oligarchy in Hawai'i, 58–60; fishing revitalization after World War II, 126–128; post–World War II fishing companies, 117–118, *118;* post–World War II rebuilding of ties with Japan, 120; seafood processing and, 63, 158; US licensing prior to World War II and, 97, 98; World War II and, 102–104, 114
otatasan (women fish peddlers), 27
otate-ura (standing beach), 17–18, 19
otoko-kasegi, Australia as place of, 31
Oyadomari, Masahiro, 136
oysters. *See* pearl oysters

pā (lure hook), 36, 171n10
Pacific Commercial Advertiser, 55
Pacific Fishing Co., 47, 52, 53–54, 89, 97, 178n67
pearl oysters, 4, 29–31, 38, 90–91
peddlers, of fish, 28, 43, 50, 58, 80, 158–160; women as, 7, 26–27, 34, 80, 81, *137,* 150
Philippines, 23–24, 27, 40, 162
"picture brides," 7, 168n11
"Plan for the Development of the Hawaiian Fisheries, A" (US Bureau of Fisheries), 99
Poindexter, Joseph B., 101
pole-and-line fishing (*ipponzuri*), 68, 155; development in Okikanuro, 16–18, 19; *tegusu* and, 15–16, 18. *See also* longline fishing

pond fishing, by Native Hawaiians, 36
Price, H. Ted, 137

Rice, Arthur H., Jr., 108, 109
Rice, Harold Frederick, 110
Robello, Lucy, 80
Russia, 4, 14
Ryukyuan-Hawaiian Brotherhood Program, 134–135

sabani (fishing boats), 130, *137,* 142
Sakashita, Genshirō, 100
Sakhalin Island, 4, 12, 14
salmon: canning of, 73, 107; fishing for, 4, 12, 28, 31–32, 55, 76, 96
sampans, 6, 48, 67, 178n; capacity and costs of, 65–66, 89; design of, 62–63; transport capacity and regulation of, prior to World War II, 92–100
Sand Island military detention camp, 102, 104
sardines, 1, 12, 21–22, 95, 144, 169n16
Satō, Torajiō, 30
seaweed, women's collection of, 24–25
Segawa, Kiyoko, 27
Seto Island Sea, 4, 15
shakushi (Japanese rice paddle), 68
Sheetz, Josef R., 133
Shiba, Sometarō, 45, 47, 57
Shigehiko, Shiramizu, 167n8
Shikoku, *13,* 17, 21
Shima-gurumi tōsō (the island-wide struggle), on Okinawa, 133
Shimizu, Haru, 72, 74, 124, *125*
Shimizu, Hisao, 105, 126, 163
Shimizu, Matsutarō, 72, 74, 104, 124–126, *125,* 163
Shimizu, Shizue, 72, 77, 96, 104–106, 124–126, 163
Shimizu, Tokiharu, 72
Shimizu, Yoshiharu, 72
Shintoism, 1, 25, 101, *118;* shrines to, 83–84, 122–124
Shinzato, Katsuichi, 131–132, 135–137, 144, 155
Shinzato, Katsumi, 146
Shinzato, Masami, 155–156
Short, Walter, 101

Shunkotsu-maru (training ship), 120, 122–123
Sino-Japanese War (1894–1895), 23
skipjack (*aku*) tuna: canning and, 63–64; early twentieth century catch, 32–33, 41; Japanese opposition to Native Hawaiian off-season for, 36–37; *katsuobushi* production and, 63; late twentieth century demand for, 155; net fishing prohibited in Hilo Bay, 56–57; Okinawans and, 131, 144–146, *146;* regulation of prior to World War II, 92–100; styles of fishing for, 66–71, *69, 70;* during World War II, 108–109, 111–112
Skipjack Tuna Fishermen's Association, 119
Smith, Andrew F., 174n82
Sōga, Yasutarō, 43–45, 47, 88, 174n83
sojourner's mentality, 5
Stainback, Ingram M., 97–98, 99
Steiner, Earnest, 101, 181n56
Steveston, Canada, 76
Suisan Co. *See* Hilo (Waiakea) Suisan
Suisan Kabushiki Kaisha, Ltd. *See* Hilo (Waiakea) Suisan
sumo tournaments, 1, 122–123, *122*
Suō-Ōshima (Yamaguchi Prefecture), 16–17, 19, 34, 42
surimi production, 63
Suzuki, Moto, 76

Tagawa, Hideo, 122
Taisho era (1912–1926), 26
Taiwan, 23, 27, 31, 42, 92, 129
Taiyo Gyogyo, 120, 127–128, 183n26
Takasagomaru (boat), 57
Takenaka, Brooks, 89
Takenaka, Isematsu, 72, 74, 89, 96, 105
Takenaka, Tokiharu, 72, 89
Tamashiro, Kiyoshi, 140, 143
Tamura, Itonosuke, 98
tanomoshi. See *kō*
Tatsumaru incident (1908), 45
Taylor, Angus, 97

Tenjinmaru (boat), 57, 101
Theo. H. Davies and Co., 6, 58–59
Thursday Island, Australia, 29–31
Tokugawa period (1603–1868), 4, 11–19, 20
Tōyama, Kyūzō, 131
training, of young fishermen, 65–66, 88–89, 120, 122–123, 134–141
trout eggs, farming of, 90–91
Tsuchiya, Nobuo, 162–163
Tsushima (island), 18–19, 21, 25, 26, 28–29
Tsushima Current, 2, 131
Tule Lake Segregation Center, 125

Ueda, Shinkichi, 53
Uehara, Hiroshige, 156
Uehara, Kameho, 184n4
Uehara, Tokusaburō, 137–138, 142, *143*, 148
ulaula (red snapper), 69–70
ulua (crevalle), 70
United Fishing Agency, Ltd., 118, 127–128, 136, 138, 140, 142, 148, 154. *See also* Honolulu Fish Auction, of United Fishing Agency
United Japanese Society of Hawaii, 134
United Okinawan Association of Hawaii (UOA), 134–137
United States, effects of annexation of Hawai'i, 39, 54–55
United States Civil Administration of the Ryukyus (USCAR), 133–137
US Army: Ōtani and, 58; US Army Pacific (USARPAC), 134, 136, 137
US Navy: Ōtani and, 59; Japanese fishing vessels and World War II, 93, 95–96, 99–101, 109–113
USS *Greeneville*, 165
utase-ami, 15, 23–24

Waiakea River, *44*
Wakayama Prefecture, 4, 5, *14;* early fishing in Kishū, 11–12, *14*, 14–16, 18, 19; fishermen from, 72, 74; fishermen in O'ahu, 64; lifestyle of fishermen from, 52–53; longline fishermen from, 70–71; travel to

Wakayama Prefecture (*continued*)
Australia from, 4, 28–31, 38; travel to Canada from, 31–32; travel to Hawai'i from, 39–41; travel to Korean peninsula from, 28
Warner, H. H., 107
Washington State, 55
West, Frank H., 109–112
whaling industry, 8, 19, 33, 169n16
Wiig, Jon, 126
Winston, E. C., 64
Wirin, A. L., 125–126
women, 6–8, 34, 75–83, *76, 79;* child care and, 9, 78, 80; early fishing, trading, and domestic activities of, 24–28; earnings of, 78; as fish peddlers, 7, 26–27, 34, 80, *81, 137,* 150; seaweed collection and, 24–25
Wood, A. A., Jr., 93
World War II: life during, 123; negotiations to reestablish fishing operations, 106–115; Okinawa during, 132; Pearl Harbor attack, martial law, and confiscation of fishing vessels, 100–106; US military presence in and aid to Okinawa after, 132–133; US suspicions about and suppression of fishing fleet prior to, 92–100
World War II, reconstruction of fishing industry after, 116–128; emotional and spiritual ties to Japan, 123–124; fishing personnel, companies, and ships, 116–119, *118, 119,* 126–127; help sought from Japan, 127–128; rebuilding of ties with Japan, 120–124; US citizenship issues, 124–126

Yabe, Gorokichi, 39
Yamaguchi Prefecture, 4, 5, 17, *17; ama* of, 25–26; Hawai'i and, 31–32, 34, 41–51; longline fishermen from, 70–71; travel to Korean peninsula from, 21–22
Yamashiro, Matsuichi, 64
Yamashiro, Matsutarō, 45, 47, *47,* 49, 53, 57, 64
Yamashiro Hotel, 45
Yamauchi, Tsuru, 77
Yanagisawa, Ikumi, 168n11
Yarnell, H. R., 95–96
Yasui, Minoru, 125
Yoshimura, Kokuichi, 85
Young, Anin, 44
Yuen, Henry S. H., 184n5

About the Author

MANAKO OGAWA is an associate professor at Ritsumeikan University in Kyoto, Japan. She received a PhD from the University of Hawai'i in American studies. While publishing articles on topics related to the Woman's Christian Temperance Union in the *Journal of World History, Diplomatic History,* and various other academic journals in both English and Japanese, her academic interests have gradually shifted to the sea and fishing communities. Her more recent articles on fishermen who traveled to Hawai'i and other parts of the Pacific have appeared in *Imin Kenkyū Nenpō, Chiiki Gyogyō Kenkyū,* and the *Hawaiian Journal of History.*

Production Notes for
Ogawa / *Sea of Opportunity*
Jacket design by Mardee Melton
Composition by Westchester Publishing Services
with display type in Galliard and text in Galliard
and Myriad Pro.
Printing and binding by Sheridan Books, Inc.
Printed on 60 lb. House White, 444 ppi.